The Internationalization of U.S. Manufacturing: Causes and Consequences

Committee for the Study of the Causes and Consequences
of the Internationalization of U.S. Manufacturing
Manufacturing Studies Board
Commission on Engineering and Technical Systems
National Research Council

NATIONAL ACADEMY PRESS
Washington, D.C. 1990

NOTICE: The project that is the subject of this report was approved by the Governing Board of the National Research Council, whose members are drawn from the councils of the National Academy of Sciences, the National Academy of Engineering, and the Institute of Medicine. The members of the panel responsible for the report were chosen for their special competence and with regard for appropriate balance.

This report has been reviewed by a group other than the authors according to procedures approved by a Report Review Committee consisting of members of the National Academy of Sciences, the National Academy of Engineering, and the Institute of Medicine.

The National Academy of Sciences is a private, nonprofit, self-perpetuating society of distinguished scholars engaged in scientific and engineering research, dedicated to the furtherance of science and technology and to their use for the general welfare. Upon the authority of the charter granted to it by the Congress in 1863, the Academy has a mandate that requires it to advise the federal government on scientific and technical matters. Dr. Frank Press is president of the National Academy of Sciences.

The National Academy of Engineering was established in 1964, under the charter of the National Academy of Sciences, as a parallel organization of outstanding engineers. It is autonomous in its administration and in the selection of its members, sharing with the National Academy of Sciences the responsibility for advising the federal government. The National Academy of Engineering also sponsors engineering programs aimed at meeting national needs, encourages education and research, and recognizes the superior achievements of engineers. Dr. Robert M. White is president of the National Academy of Engineering.

The Institute of Medicine was established in 1970 by the National Academy of Sciences to secure the services of eminent members of appropriate professions in the examination of policy matters pertaining to the health of the public. The Institute acts under the responsibility given to the National Academy of Sciences by its congressional charter to be an adviser to the federal government and, upon its own initiative, to identify issues of medical care, research, and education. Dr. Samuel O. Thier is president of the Institute of Medicine.

The National Research Council was organized by the National Academy of Sciences in 1916 to associate the broad community of science and technology with the Academy's purposes of furthering knowledge and advising the federal government. Functioning in accordance with general policies determined by the Academy, the Council has become the principal operating agency of both the National Academy of Sciences and the National Academy of Engineering in providing services to the government, the public, and the scientific and engineering communities. The Council is administered jointly by both Academies and the Institute of Medicine. Dr. Frank Press and Dr. Robert M. White are chairman and vice-chairman, respectively, of the National Research Council.

This study was supported by the Federal Emergency Management Administration and the National Science Foundation under Contract No. DMC-871-3483 between the National Science Foundation and the National Academy of Sciences, and by the Academy-Industry Program of the National Research Council.

Library of Congress Catalog Card Number 90-62810
International Standard Book Number 0-309-04331-X

Limited copies are available from:

    Manufacturing Studies Board
    National Research Council
    2101 Constitution Avenue, HA 270
    Washington, DC 20418

Additional copies are available for sale from:

    National Academy Press
    2101 Constitution Avenue, N.W.
    Washington, D.C. 20418

S214

Printed in the United States of America

COMMITTEE FOR THE STUDY OF THE CAUSES AND CONSEQUENCES OF THE INTERNATIONALIZATION OF U.S. MANUFACTURING

PAUL J. KEHOE (*Chairman*), Vice Chairman (retired), Kellogg Company, Battle Creek, Michigan
CLAUDE E. BARFIELD, Director, Science and Technology Studies, American Enterprise Institute, Washington, D.C.
KAN CHEN, Professor of Electrical Engineering and Computer Science, University of Michigan, Ann Arbor
CHARLES E. EBERLE, Executive Vice President, Consumer Products Business, James River Corporation, Richmond, Virginia
MURRAY FINLEY, President (retired), Amalgamated Clothing and Textile Workers Union, New York, New York
HERBERT I. FUSFELD, Director, Center for Science and Technology Policy, School of Management, Rensselaer Polytechnic Institute, Troy, New York
HOWARD K. GRUENSPECHT, Economic Advisor to the Chairman, International Trade Commission, Washington, D.C.
IAN HANCOCK, Managing Director, Putnam, Hayes & Bartlett, Ltd., London, England
WILLIAM C. HITTINGER, Executive Vice President (retired), RCA Corporation, Summit, New Jersey
WILLIAM G. HOWARD, JR., National Academy of Engineering Senior Fellow and Senior Vice President and Director of R&D, Motorola, Inc. (on sabbatical), Scottsdale, Arizona
MELVIN KUPPERMAN, President, A. Epstein & Sons International, Inc., Chicago, Illinois
TINA M. MARQUEZ, Purchasing Manager, Apple Computer, Inc., Fremont, California
DAVID C. MOWERY, School of Business, University of California, Berkeley
WILSON NOLEN, Corporate Vice President and Assistant to the Chairman and CEO, Becton Dickinson and Company, Franklin Lakes, New Jersey
MICHAEL OPPENHEIMER, Executive Vice President, The Futures Group, Glastonbury, Connecticut
C. K. PRAHALAD, Professor, Corporate Strategy and International Business, School of Business Administration, University of Michigan, Ann Arbor
MICHAEL RADNOR, Professor, Kellogg Graduate School of Management, and Director, Center for Interdisciplinary Study of Science and Technology, Northwestern University, Evanston, Illinois

JOHN C. READ, Vice President & General Manager, Engine Group, Donaldson and Company, Minneapolis, Minnesota

J. RONALD STEIGER, JR., Vice President for Computer-Integrated Manufacturing, IBM Corporation, Purchase, New York

W. EDWARD STEINMUELLER, Deputy Director, Center for Economic Policy Research, Stanford University, Stanford, California

SIDNEY TOPOL, Chairman of the Board, Scientific-Atlanta, Inc., Atlanta, Georgia

Staff

THOMAS C. MAHONEY, Project Director, and Acting Director of the Manufacturing Studies Board

KERSTIN B. POLLACK, Deputy Director of the Manufacturing Studies Board, and Director of New Program Development

ERIC A. THACKER, Research Associate

LUCY V. FUSCO, Staff Assistant

MANUFACTURING STUDIES BOARD

JAMES F. LARDNER (*Chairman*), Vice President (retired), Component Group, Deere & Company
MATTHEW O. DIGGS, JR., Vice Chairman, Copeland Corporation
GEORGE C. EADS, Vice President, Product Planning and Economics, General Motors Corporation
HEINZ K. FRIDRICH, Vice President, Manufacturing, IBM Corporation
LEONARD A. HARVEY, Executive Vice President (retired), Borg-Warner Chemical Company
EDWARD E. LAWLER III, Director, Center for Effective Organization, University of Southern California
JOEL MOSES, Head, Department of Electrical and Computer Engineering, Massachusetts Institute of Technology
LAURENCE C. SEIFERT, Vice President, Communications and Computer Products, Sourcing and Manufacturing, AT&T
JOHN M. STEWART, Director, McKinsey and Company, Inc.
WILLIAM J. USERY, JR., President, Bill Usery Associates, Inc.
HERBERT B. VOELCKER, Charles Lake Professor of Engineering, Sibley School of Mechanical Engineering, Cornell University

Preface and Acknowledgments

Confronted with ever-increasing volume of foreign products competing for domestic and global market share, a rapidly rising number of foreign companies establishing manufacturing operations in the United States, and world-wide dispersion of skills and technology, U.S. manufacturing finds itself in a new, largely unfamiliar, competitive environment. Global competition has become a powerful driving force behind manufacturing investment, operations, and strategic decisions.

Of course, U.S. multinationals have led the way in foreign investment, building global manufacturing presence and gaining global market share by internationalizing their operations. However, the pace of change in global markets has accelerated in the last 15 years, with unprecedented levels of penetration of the U.S. market through imports and direct investment, growing competitive challenges to American products in foreign markets, and a rise in the size, number, and capabilities of foreign multinationals that has eliminated the dominance of U.S. firms. Internationalization has forced rapid change on companies historically immune to foreign competition and, in a short time, totally redefined the meaning of competitive manufacturing.

The pace of change has arguably left many companies unprepared. Accustomed to serving domestic customers and fighting well-known competitors, many firms have had difficulty adapting to new competition. The need to help U.S. manufacturers and policymakers respond to greater foreign competition and continued international interdependence led directly to this project. With funding from the Federal Emergency Management

Agency, the National Science Foundation, and the Academy-Industry Program, the Manufacturing Studies Board of the National Research Council formed the Committee on the Causes and Consequences of the Internationalization of U.S. Manufacturing. The committee was asked to examine the responses of U.S. manufacturers to trends in international competition and to relate these competitive responses to current and prospective government policies.

The findings and analysis contained in this report are based on the committee members' experience either managing, studying, or advising major multinational manufacturers. Information was gathered through interviews with senior managers from manufacturing companies in industries as diverse as biotechnology, paper products, and auto parts. In addition, professors from the Center for the Study of U.S.-Japan Relations at Northwestern University, led by Dr. Atul Wad, conducted interviews for the committee with senior manufacturing managers in Japan. The report has also benefitted from a parallel effort by a National Academy of Engineering study committee that has explored the globalization of technology and its policy implications for the United States (*National Interests in an Age of Global Technology*, 1990).

Based on its discussions and analysis of the current environment for international competition, the committee has written this report to dispel misconceptions regarding the drivers of internationalization and, therefore, to improve understanding of both the challenges and the opportunities of a global market and production base. Important consequences of internationalization for both manufacturers and national policy are described. Finally, the committee provides its assessment of what it takes to be successful as manufacturers and as a nation in the international competitive environment.

The Committee on the Causes and Consequences of the Internationalization of U.S. Manufacturing is responsible for organizing and conducting the research and writing the findings of this study. Our work would not have been possible without the contributions of the Manufacturing Studies Board staff: former executive director George Kuper, deputy director Kerstin Pollack, senior staff officer Tom Mahoney, and administrative assistant Lucy Fusco. We also wish to thank Proctor Reid for his assistance during the early stages of the project and Kenneth Reese for his help in editing the final report.

Paul J. Kehoe
Chairman

Contents

EXECUTIVE SUMMARY 1

1. INTRODUCTION 5

2. CAUSES OF INTERNATIONALIZATION 9

    Changes in Global Markets, 10
    Global Dissemination of Technology, 16
    Changes in Cost Priorities, 22
    Political and Economic Factors, 26
    Conclusion, 30

3. NATIONAL ECONOMIC AND POLITICAL IMPLICATIONS .. 34

    Inward Investment and Foreign Ownership, 34
    Technology Flows, 37
    Domestic Versus International Policy, 38
    Inadequate Information, 39
    Conclusion, 40

4. KEYS TO SUCCESS 41

    Sources of Corporate Success, 41
    Sources of National Success, 50

5. CONCLUSION 57

APPENDIX: INDICATORS OF INTERNATIONALIZATION ... 59

BIBLIOGRAPHY 63

The Internationalization of U.S. Manufacturing: Causes and Consequences

Executive Summary

Internationalization is an increasingly pervasive force in U.S. manufacturing, creating new sources of competition and new standards for competitiveness. The growing importance of imports and exports in domestic manufacturing and the significant rise in foreign investment in the United States in recent years are the most obvious evidence of internationalization. Less obvious, but more important, are the interdependent relationships being established across national borders. International networks of suppliers, customers, researchers, technology developers, and distributors have emerged, creating an unprecedented degree of global interdependence. The formation of these networks is being driven by manufacturing managers seeking to maximize competitive advantages in response to changes in markets, costs, technologies, and politics.

Markets are global. Manufacturers cannot afford to ignore the revenue potential of foreign markets, the necessity of attacking competitors abroad to protect domestic market share, or the advantages of learning the demands of customers in diverse markets. For many product lines, penetration of global markets depends on having local production capacity for quick response to customer demand and manufacturing systems that can achieve constant improvement in cost, quality, and value.

Cost priorities have shifted. Cheap labor no longer dominates decisions to manufacture offshore; few industries have product lines with sufficient labor content to justify investment strategies based solely on labor costs. Manufacturing costs are being driven by process control and flexibility, product quality, customer responsiveness, and the skills needed to spur constant improvement. The need to control total system costs determines the resources sought in decisions on international investment locations as well as the level of sophistication to be used in foreign plants.

Technology is global. The sources of new technology have multiplied, and U.S. dominance in the creation of new technology has ended. Many countries have the human resources and scientific infrastructure to excel at research and technology development. Competitive success depends on how quickly and effectively new technology, from whatever source, is incorporated into new products and processes.

The international economic and political environment is increasingly complex. Flexible exchange rates, the magnitude of global capital flows, and the virtual elimination of tariffs among developed countries, combined with the emergence of nontariff barriers as the dominant form of protectionism, have opened new market opportunities and introduced new sources of risk for global manufacturers. Internationalization has clouded the distinction between domestic and foreign policy and created pressure for greater international cooperation in setting national policies.

Too many domestic manufacturers continue to base their strategies on the U.S. market, U.S. competitors, and U.S. technology. The changes wrought by internationalization are well understood by large multinational companies, both in the United States and abroad. In fact, U.S. multinationals have a more extensive global production base and longer experience at managing that base than firms of other nations, potentially creating a source of competitive advantage as internationalization progresses. On the other hand, the size of the U.S. market has long insulated domestic manufacturers from the pressures of foreign competition. Too few U.S. manufacturers yet realize the importance or the pervasiveness of the internationalization process. They *react* to foreign competition in this market, rather than building a strategy for attacking foreign markets, gaining access to foreign technology, and building world-class production capabilities.

Similarly, national policies must be placed in the context of global competition and global dissemination of technology. Internationalization has introduced considerations for policy that were less important or nonexistent when the United States was virtually self-sufficient. For instance, the need to monitor and gain access to foreign technological developments has emerged as a major national priority. Similarly, the growth of foreign ownership of the domestic production base and the global dissemination of U.S.-owned production facilities have blurred the definition of a company as American or foreign, bringing new complications to established policies in areas such as trade and technology. Above all, internationalization has created a greater degree of choice for manufacturers pursuing global market share and profitability objectives, thereby generating greater competition among nations for manufacturing investment. The United States cannot rely on market size alone to attract the high-value manufacturing activities that are essential for building national wealth. Manufacturers will place their high-value activities where the resources, skills, and infrastructure are available to perform them most effectively. This fact dictates the standards that must be met in maintaining a favorable environment for investment, innovation, and operations.

Given these realities, what factors are essential for corporate and national success in the 1990s? For both, internationalization must become the cornerstone of policy. The knowledge that global competition will be intense and constant, with different participants striving to maximize unique competitive advantages, must be the foundation of strategy. The dynamism of such a competitive environment dictates corporate and national strategies that are equally dynamic. The ability to implement these strategies in turn requires manufacturing resources that allow flexibility, constant improvement, and a focus on innovation.

For manufacturers the essential ingredients of a successful strategy include:

- Developing managers with a broad understanding of not only foreign markets and international competitors but also the technology and the knowledge to use it needed by world-class manufacturers.
- Building corporate intelligence on international markets, technological developments, competitors' strategies, consumer demand, and political changes as a prerequisite for aggressive pursuit of global market share.
- Strengthening core competitive capabilities—the combination of technologies, skills, and products that provide competitive advantages that are difficult for competitors to imitate or overcome.
- Using international relationships with suppliers, researchers, technology developers, and producers to leverage in-house resources while also using and absorbing outside expertise to strengthen core competitive capabilities.
- Speeding commercialization of new technologies by improving access to and utilization of external technology, integrating product and process design and engineering, and emphasizing rapid time to market as a core competence.

Similar considerations apply in creating national policies that will strengthen U.S. competitive advantage in an environment of national competition for manufacturing investments. To ensure that the United States remains an attractive location for high-value manufacturing activities, policymakers must take steps to:

- Spur competition in the United States by continuing to resist protectionist pressures and, at the same time, work to ensure that U.S. companies have open access to global markets.
- Build the intellectual assets required to perform advanced manufacturing functions by investing in an effective educational infrastructure with emphasis on constantly upgrading the skills of the existing work force and on imparting state-of-the-art engineering and technology management practices to new graduates.
- Reassess the national information requirements for understanding the changing position of the U.S. economy and U.S. corporations in the global environment, thereby facilitating creation and implementation of appropriate national policies.
- Emphasize high-value manufacturing as a national core competitive capability by providing the incentives and resources—intellectual, capital, infrastructural—required to encourage private manufacturers to perform high-value activities in the United States.

Internationalization has already made these factors essential to corporate and national success in the global economy. Unfortunately, U.S. corporations will find no generalizable prescriptions for acting on them. Each firm must match itself against its global competitors to establish benchmarks, assess market and technology potential on a global basis, and develop an implementable strategy based on its unique capabilities and objectives. Similar steps are needed on a national level. Little effective national strategy can be expected until the United States establishes her position vis-à-vis her global competitors; strengths must be maximized and weaknesses corrected. Such national action is the only way to implement a proactive strategy that builds on U.S. successes rather than reacting to foreign achievements.

To maintain leadership in the dynamic world created by internationalization, the nation must recognize that global interdependence is here to stay. Such interdependence must be managed, through cooperation and collaboration between industry and government, including foreign partners, both to ensure continued advances in the global manufacturing system and to maximize U.S. interests as this global system evolves. The required actions demand a national consensus that international manufacturing competitiveness deserves the highest national priority and, therefore, should command whatever resources are necessary.

1
Introduction

The 1980s witnessed dramatic changes in the world's manufacturing environment, creating new bases for competitive advantage and opportunities for new participants to compete internationally. Some of these changes are driven by technical advances, such as the application of new materials and electronic devices to products and processes and the use of information systems to integrate manufacturing functions. Others are due to advances in organizational and management practices in manufacturing systems designed to achieve total quality control and production flexibility. Still other changes are driven by economic factors, such as the relative parity in income, consumer demand, and educational levels among developed countries; the growing wealth of newly industrialized countries (NICs); and the unprecedented level of macroeconomic imbalances among major trading nations. Manufacturers are adapting to these changes by building networks of production, engineering, and research and development (R&D) capabilities, frequently in partnership with other firms, to establish market presence, generate profits, gain access to technology, leverage resources, and enhance flexibility on a global basis.

The responses of individual companies seeking to maintain competitive advantage in this new environment are internationalizing the U.S. manufacturing base more pervasively than historical investment flows.[1] Although U.S. manufacturers became multinational earlier than most foreign

---

[1] The bibliography lists a number of sources of information on the historical development of multinational corporations and internationalization.

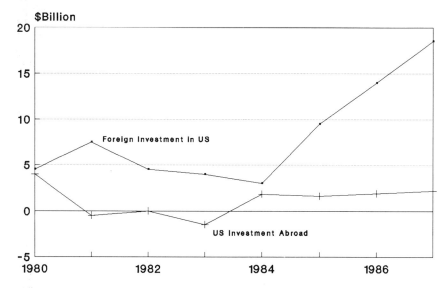

FIGURE 1 New international investment in manufacturing, 1980-1987. Source: Bureau of Economic Analysis.

firms—Singer, for example, began producing sewing machines in Europe in the 1860s—the flow of new manufacturing investment abroad peaked by the early 1970s and was negligible in the 1980s (see Figure 1).[2] At the same time, the flow of foreign investment into the United States has risen dramatically, led by the United Kingdom with 31 percent, Japan with 16 percent, and the Netherlands with 15 percent.[3] This change in the direction of capital flows, the installed global manufacturing base of U.S. multinationals, the rapid increase in international trade, and the growing parity in global technological capabilities have fostered a degree of international interdependence unknown in the past. (The Appendix presents a number of economic indicators demonstrating the extent of internationalization in the U.S. manufacturing sector.)

In the modern context, internationalization can be defined as

a dynamic process of developing, leveraging, and managing manufacturing, marketing, and R&D activities to achieve business objectives in a globally competitive environment.

The implications of this dynamic process for U.S. manufacturers are

---

[2] Stephen Cooney, *Manufacturing Creates America's Strength*, Washington, D.C., National Association of Manufacturers, December 1988, p. 26.

[3] These percentages indicate the distribution of the stock of foreign direct investment in the United States in 1988. See Edward M. Graham and Paul R. Krugman, *Foreign Direct Investment in the United States*, Washington, D.C., Institute for International Economics, 1989, p. 34.

profound. Because of the strength and pervasiveness of foreign competition, few manufacturers can afford to focus only on the U.S. market. Small supplier firms with a domestic orientation may need to rely on their customers to provide the knowledge of foreign capabilities and secure the markets necessary to remain competitive against foreign competition. Small and medium-sized firms will need to expand their global presence, initially through exports, to build production volume, tap changing consumer tastes and technological developments, and engage their competitors to preempt the advantages they gain by dominating large overseas markets. Large multinational firms face the challenge of integrating their international activities to maximize cross fertilization of ideas and to minimize transaction costs. Finally, high-technology firms in industries such as electronics, materials, and aerospace face an additional challenge. As the costs of R&D and capital equipment rise, relatively few firms will have the capital resources and technological capabilities in-house to act independently; managers must consider an array of options, including joint ventures, technology development partnerships, selected relationships with original equipment manufacturers, and other forms of alliances. Regardless of company size or industry, the ability to participate in global networks of suppliers, distributors, and researchers while retaining unique competitive capabilities has become an essential determinant of corporate success. This imperative distinguishes the current international environment from its historic predecessors.

The emerging internationalization process, by creating global manufacturing systems and interlocking networks of firms in both formal and informal relationships, generates a degree of interdependence that, while mutually advantageous for the participating companies and countries, creates invisible risks and introduces new management challenges that are often poorly understood. Policymakers in both industry and government need to recognize the extent and inevitability of the internationalization process, the mechanisms used to create competitive advantage in response to it, and the repercussions for both corporate and national policy.

The purpose of this report is to foster the necessary understanding among decision makers. The findings reflect the corporate and academic experiences of the members of the committee, augmented by a review of the relevant literature and interviews with senior managers from diverse manufacturing industries, including food processing, microelectronics, automobiles and parts, consumer products, biotechnology, and precision instruments. The diversity of experiences in grappling with these issues made consensus difficult to achieve but revealed a number of general themes regarding the causes and consequences of internationalization.

The report first identifies the primary forces now driving manufacturers' international business behavior—the causes of internationalization. Next it describes the consequences of private decisions for the national economic and policymaking environment. Because of the unique set of skills and resources each manufacturer brings to its international activities, no attempt is made to generalize the consequences for private business. Finally, the report outlines fundamental steps necessary for both private and national success. These "keys to success" are based on the committee's central conclusion that manufacturers must act in their own best interests and, therefore, must be able both to define their interests accurately and to act accordingly. Policymakers in turn must strive to provide a national economic environment in which corporations, acting in their own best interests, will benefit the nation's well-being.

2
Causes of Internationalization

Although U.S. manufacturing has been international in scope for most of this century, the past 10 to 15 years have witnessed the evolution of a new global manufacturing environment. Previously, the international market was the preserve of large multinational corporations and generally was ignored by domestic firms. Today it is essential that virtually all manufacturers be aware of and participate in international markets. Reasons for the increased importance of internationalization include the following:

- Foreign competition in the U.S. market has grown incessantly.
- Rapid increases in foreign demand for manufactured products have created highly attractive markets abroad.
- The pace of technological development has accelerated as sources of new technology have diffused worldwide.
- Changes in global political and economic conditions have created a variety of new opportunities and challenges for manufacturers.

These factors have created not only a new global market for manufacturing inputs and products but also a need for international awareness unknown in previous eras. They have also opened an array of options for participating in international markets that were unavailable in the past. Each of these factors affects different industries, even different products, in varying degrees; generalizations are impossible in a sector as diverse as manufacturing. Closer examination will reveal some of the differences and provide a better understanding of the internationalization process.

CHANGES IN GLOBAL MARKETS

The nature of global markets and the role of the United States in the world marketplace have shifted fundamentally since the mid-1970s. Relevant changes range from broad economic trends that have altered the relative importance of various national and regional markets to technological developments that have changed the criteria for competitive success in those markets.

Foreign Competition in the United States

One of the most important developments is the increasing openness of the U.S. economy. Import penetration in consumer goods grew from about 7 percent of domestic consumption in the early 1980s to more than 11 percent in 1988 (see Figure 2). In capital goods, penetration rose from about 14 percent to almost 40 percent in the same period (see Figure 3).[1] The rapid increase in foreign-owned production capacity in the United States further expands foreign competition in the domestic market. Assets of foreign manufacturing affiliates as a percentage of total assets of U.S. manufacturing corporations rose from 5.2 percent in 1977 to 12.2 percent in 1987; similarly, the number of employees working for foreign manufacturing affiliates as a percentage of total U.S. employment more than doubled from 3.5 percent in 1977 to 7.9 percent in 1987 (see Figure 4).[2] Foreign competition has become so extensive in virtually every industry that all companies, even small- and medium-sized firms that have not historically produced or marketed products abroad, must contend with foreign firms that are either direct competitors in the United States or have the potential to be.

The pervasiveness of foreign competition is forcing domestic manufacturers to upgrade their manufacturing operations or face serious competitive difficulties. Achieving a combination of cost, quality, delivery, and performance that maximizes value to the customer has become the determinant of global manufacturing success. Just-in-time and total quality control systems have become the minimum ante for world-class manufacturing operations. Other measures, such as improving productivity by enhancing workers' skills and encouraging their participation in decision making, strengthening customer-supplier relations to instill a sense of partnership, and effectively implementing advanced manufacturing technologies to enhance flexibility and spur innovation, are becoming essential to competitive manufacturing

---

[1] Rudiger Dornbusch, James Poterba, and Lawrence Summers, *The Case for Manufacturing in America's Future*, Rochester, N.Y., Eastman Kodak, 1988, and Federal Reserve Board, unpublished data, 1989.

[2] Graham and Krugman, *Foreign Direct Investment in the United States*, p. 13.

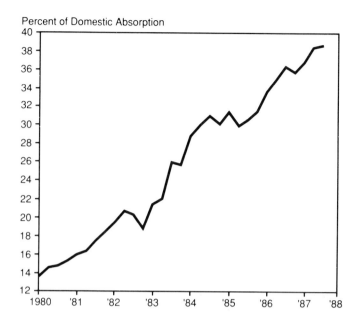

FIGURE 2 Import penetration—consumer goods.

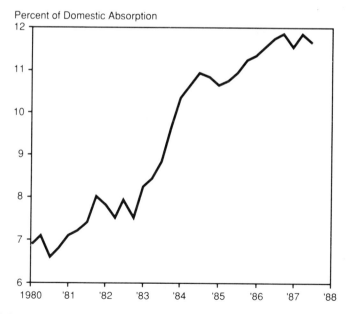

FIGURE 3 Import penetration—capital goods.

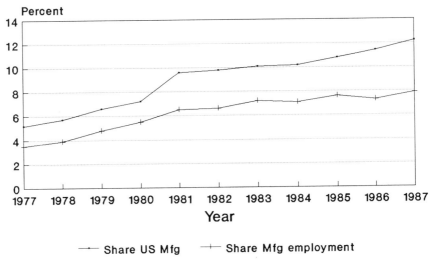

FIGURE 4 Foreign role in U.S. manufacturing, 1977-1987 (percentages). Source: Graham & Krugman.

in the domestic market. The need to stay abreast of product and process innovations worldwide and to build genuinely world-class operations is being forced on companies of all sizes, even those with no historical international presence.[3]

Growth in Foreign Demand

The second major development in the global market is the shift in the relative size of markets in the United States and abroad. In 1965 the United States accounted for 40 percent of world gross domestic product; by 1987 the U.S. share percentage had fallen to about 30 percent (see Figure 5). Over the same period, growth in private consumption abroad exceeded the rate in the United States.[4] In particular, rapid economic growth in developing countries has greatly expanded the number, sophistication, and wealth of markets around the world. In the past, U.S. manufacturers could be satisfied with the domestic market or at least confident that the United States was their leading market, but many are now finding that an increasing proportion of their current sales and a large proportion of

---

[3] A number of authors have addressed the need to upgrade manufacturing capabilities to be globally competitive. See, for example, Kim Clark, Robert Hayes, and Steven Wheelwright, *Dynamic Manufacturing*, New York, The Free Press, 1988.

[4] The World Bank, *World Development Report 1989*, New York, Oxford University Press, 1989, pp. 169, 179.

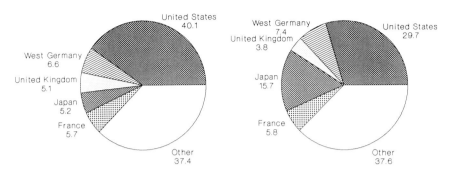

World Bank World Development Report 1989

FIGURE 5 Share of world GDP 1965, 1987.

potential future demand are abroad. Those firms that have aggressively pursued foreign markets have also discovered the benefits of a global presence in smoothing demand fluctuations, creating significant advantages over competitors concentrated in one market.

In its discussions with manufacturing executives, the committee found a growing realization that simply exporting to foreign markets may not provide the broad access, interaction with customers, and learning opportunities needed for long-term growth in global market share. Manufacturing managers often cite this realization as the primary motivation for siting production facilities abroad.

There are a number of reasons. One is the advantages of bringing global assets and capabilities to bear on local markets. For instance, a U.S. medical products firm with a global manufacturing and marketing capability has used integrated global production and R&D to compete successfully with a host of small competitors in protected local markets. Another reason is the advantages of encountering a range of strategies used by multiple competitors. The lessons learned help to identify the critical success factors that apply in any market. Finally, a global presence helps guard against complacency and gives the ability to retaliate against competitors in their home markets. By facing challenges in foreign markets, U.S. firms can preempt competitors' ability to use a strong foreign market position as a lever for aggressive marketing in the United States. For instance, the strength of the Kellogg Company in global markets for ready-to-eat cereals has been a major reason why large foreign competitors such as Nestle have a limited presence in the U.S. cereal market. A company that is unable to retaliate against foreign competitors that enter its domestic market is at a serious disadvantage.

Shorter Product Lives, More Customization, Faster Response

Advanced design and manufacturing technology and management practices have given companies in many industries the ability to introduce new products more rapidly and to customize products to define and attack a range of market niches. The result is shorter product life cycles. In microelectronics, for example, the number of functions on a dynamic random access memory (DRAM) chip has doubled every 2 years, creating a new, denser commercial chip every 3 to 4 years;[5] the resulting improvements in memory then drive next-generation products in computers, consumer electronics, and a number of other industries as well as demanding rapid improvements in processing equipment. In automobiles the Japanese have compressed vehicle development lead times to less than 4 years, allowing more frequent model changes and better response to customer demand; U.S. and European firms are striving to catch up. The trend is also apparent in industrial goods markets. For instance, Allen-Bradley uses advanced manufacturing technologies to produce customized contactors and relays in batches as small as one; General Electric does the same with circuit breakers. As the technology advances and permeates more companies, rapid response time and customization will become competitive necessities.

The implications of this change for corporate and government strategy are becoming apparent. Only a few industries remain in which the traditional product cycle approach to international production still applies.[6] Strong intellectual property rights have become more important as companies are forced to recoup investments on new products and production facilities over shorter periods. Manufacturers cannot afford the revenue losses associated with misappropriation of intellectual property. The International Trade Commission estimates that inadequate protection of intellectual property cost U.S. firms $24 billion in lost sales in 1986.[7] Finally, possession of the resources, expertise, and production capabilities needed to participate in multiplying market segments with rapid product turnover is becoming a key element in corporate and national competitive success. To meet international competition, manufacturers must be able to organize frequent product launches on a global basis, which increasingly requires multinational manufacturing and engineering capabilities. For example,

---

[5] Thomas Howell, William Noellert, Janet MacLaughlin, and Alan Wm. Wolff, *The Microelectronics Race: The Impact of Government Policy on International Competition*, Boulder, Colo., Westview Press, 1988, p. 38.

[6] For a fuller description of many of the issues related to trade and technology, see Raymond Vernon, "Coping with Technological Change: U.S. Problems and Prospects," in Harvey Brooks and Bruce Guile, eds., *Technology and Global Industry*, Washington, D.C., National Academy Press, 1987.

[7] "The Sun Also Rises over Japan's Technology," *The Economist*, April 1, 1989, p. 57.

Apple Computer has built an international manufacturing and engineering infrastructure with facilities in California, Ireland, and Singapore, to allow broad, simultaneous product introduction in all of its major markets.

Increasing Capital Intensity

A number of high-technology industries have experienced escalations in the capital intensity of production facilities that forced them to adopt a global marketing strategy combined with a concentrated manufacturing base. For example, in microelectronics, greater capital intensity, expensive product and process R&D, and escalating costs of capital equipment have made the ability to sell products globally essential to achieving the requisite scale economies for amortizing investments. In 1986 facilities' charges for state-of-the-art semiconductor plants ranged between $50 million and $100 million; recent estimates have priced new plants at $250 million to $400 million. When R&D costs are included, estimates for 16-megabit DRAM facilities exceed $1 billion. This level of capital intensity has two major effects on corporate strategy. First, it forces manufacturers to share the costs and the risks. For example, Texas Instruments and Hitachi, Motorola and Toshiba, and IBM and Siemens have reached agreements for joint work in developing 16-megabit DRAMs and the requisite process technology. High capital requirements also demand that costs be amortized by maximizing capacity utilization, moving rapidly along the learning curve to boost production yields, and attacking markets on a global scale. This demand implies a strategy of concentrating production in a few locations and exporting. These forces help to explain the importance of open markets in semiconductors, as demonstrated by the U.S. demand for greater access to the Japanese semiconductor market.

Market Sophistication

In a variety of industries a presence in specific markets abroad has become essential to maintaining a technological edge. For historical and other reasons, certain markets have developed an industry mix that demands state-of-the-art products in certain areas. In these state-of-the-art markets, constant improvement in product and process technologies is driven by customer demand; therefore, successful innovation becomes a key determinant of competitive success. Participating in these markets is essential both to keep pace with technological advances in a given product class and to provide ready access to a customer base for innovations that can ease the risks and build a basis for global sales. Japan, for instance, has become the state-of-the-art market for semiconductor process equipment and consumer electronics. Germany and Japan probably share this distinction in machine tools. The United States can still claim to be the leading market in, for example, aerospace, computers, and software technologies, but in many technologies crucial to manufacturing (process equipment) or to capturing large markets for manufactured products (consumer electronics), *the United States is not the state-of-the-art market.*

Firms competing on the basis of the technological sophistication of their products have no choice but to build a strong presence in these state-of-the-art markets. Effective participation through exports alone, though not impossible, is very difficult. Production capabilities in the market, along with significant engineering resources, are essential to identifying and responding to rapidly changing consumer demand or competitors' challenges. The lessons learned can then be transferred to the firm's other production facilities worldwide and used to gain a competitive edge in world markets.

These examples illustrate the kinds of issues that the globalization of markets has introduced to manufacturing strategies. Operational implementation takes a range of forms, depending on the industry, product sophistication and maturity, size and resources of specific companies, political constraints, and other factors. Conditions might dictate a wholly owned investment strategy or, increasingly, some form of collaboration with a firm that already has a market presence. Managers at one auto parts company explained that joint ventures are now its favored approach to building international market share in mature product markets, but the firm still depends on direct investment to build a presence in new businesses. In particular, many U.S. manufacturers view joint ventures as the most effective way to penetrate the Japanese market; such alliances can help overcome close manufacturer-supplier relations and language barriers.

GLOBAL DISSEMINATION OF TECHNOLOGY

Another major factor in the globalization process is the development of technological strengths in many foreign firms and economies that are overtaking the United States in a number of critical areas. Various indicators support this contention. Japan, Germany, and France devote a greater percentage of gross national product (GNP) to nondefense R&D than the United States. By 1988 the number of U.S. patents granted to foreign inventors, primarily from Japan, Germany, France, and Britain, virtually matched those to U.S. inventors (see Figure 6). Finally, the share of the U.S. market for high-technology goods supplied by imports climbed from a negligible 5 percent in 1970 to 18.2 percent in 1986; over that period the sources of such imports expanded beyond Europe to include

Japan (the dominant supplier) and the Asian newly industrialized countries (NICs—Hong Kong, Singapore, South Korea, and Taiwan).[8]

A key ramification of this diffusion of technological capability is the need for U.S. manufacturers to improve their ability to tap multiple sources of technology and to absorb new technologies into their products and processes. Despite many exceptions, certain evidence shows that U.S. manufacturers are handicapped in global competition by their poor ability to absorb new technologies, particularly from external sources, that form the basis for building new competitive advantages and commercializing new products rapidly. A recent study of the time and funds needed by Japanese and U.S. firms to commercialize new technology showed U.S. manufacturers to be at a clear disadvantage in commercializing external technology, though for internally developed technology the two countries were about even. Although this disadvantage may be due to differences in the way resources are allocated in the innovation process, it is also an indication of the costs of a pervasive "not-invented-here" attitude in U.S. industry that must be overcome.[9] The global dissemination of technology, both hardware and "soft" management systems, has made this shortcoming of U.S. manufacturers a major detriment to their global competitiveness.

Despite the evidence that U.S. companies do not integrate new technologies into their operations as well as many foreign competitors, an increasing number of firms are recognizing the importance of tracking and gaining access to technological developments worldwide. As with the other drivers of the internationalization process, the steps needed to do so cannot be generalized. They depend on the firm's own technological assets, whether competitiveness is based on product or process technology or both, whether the firm strives to be a technological leader or a quick follower, and the relative availability of foreign technology. Techniques used to gain access include creating wholly owned R&D facilities in key foreign markets, contracting with independent research institutions, strengthening ties to local universities, establishing local production facilities in areas with relevant technological concentrations, entering joint ventures with foreign firms in the United States or abroad, buying key technologies (embodied in products or processes) from foreign suppliers, licensing foreign patents, reviewing local technical publications, and informal cooperative research with foreign companies. A few examples will help to illustrate the kinds of steps that manufacturers are taking to ensure global access to technology.

---

[8] These and other statistics on R&D, technology trade, and production are available in National Science Board, *Science and Engineering Indicators—1989*, Washington, D.C., U.S. Government Printing Office, 1989.

[9] Edwin Mansfield, "Industrial Innovation in Japan and the United States," *Science*, September 30, 1988, p. 1769.

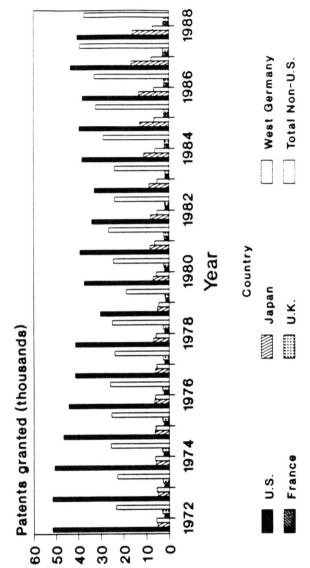

FIGURE 6 Annual share of U.S. patents 1970-1988 by nationality of inventor. Source: National Science Board, *Science and Engineering Indicators—1989*, Washington, D.C., U.S. Government Printing Office, 1989.

Access to Critical Components

Perhaps the most common way to gain access to technology is to purchase components from outside suppliers. Sourcing decisions usually involve a sophisticated analysis of which products and components should be manufactured in-house and what can be bought from suppliers. The optimum balance between make and buy depends on a variety of factors, not just access to technology; they include cost, supply flexibility, necessary process capabilities, and the role of specific components or products in overall corporate strategy. In some product segments, however, buying components from foreign manufacturers has become the only way to participate in final product markets.[10] For instance, Canon dominates global production of engines for facsimile machines and laser printers, with 84 percent of the world market.[11] Similarly, Fanuc holds more than 70 percent of the global market for machine tool controllers.[12] These market shares imply not only monopolistic power over price and deliveries but also virtual control of the pace of innovation in the affected product markets. The customers' need to guarantee access to such critical components can be the determining factor in corporate decisions ranging from plant location to frequency of model changes. In many cases, heavy dependence on single suppliers for critical components can be reversed with appropriate investments—IBM, for instance, manufactures its own laser printer engines—but the ability to build the necessary production capabilities with the requisite skill and knowledge becomes more difficult with each new generation of component technology.

Access to Process Equipment

Competitive product technologies often depend on the most advanced process technologies, which in some industries requires that process equipment be obtained from foreign suppliers. For example, U.S. semiconductor manufacturers depend on Japanese suppliers for certain advanced materials (glass, ceramics, and some specialized chemicals) and increasingly for leading-edge lithographic equipment. U.S. auto companies now must import state-of-the-art machine tools in some applications to achieve world-class parts quality. The rising level of import penetration in the U.S. capital equipment market in the 1980s is a clear indication of the overall strength of foreign suppliers (see Figure 3).

---

[10] For a discussion of U.S. dependence on Japanese suppliers of microelectronic components, see Robert B. Reich, "The Rise of Techno-Nationalism," *The Atlantic Monthly*, May 1987, pp. 63-69.

[11] "Canon," *Financial Times Supplement*, April 27, 1988, p. 16B.

[12] Gene Bylinsky, "Japan's Robot King Wins Again," *Fortune*, May 25, 1987, p. 53.

In most cases, however, it is not enough to buy the equipment. Access to the skill needed to operate it effectively also may be necessary, requiring the establishment of training facilities or manufacturing plants in foreign countries. For example, both IBM and Xerox have recently decided to produce their next generation of video displays in Japan, which has the best process capability for this technology. One large U.S. paper company has been building operations in Europe not only to penetrate the large European market before 1992 but also to gain timely access to process developments emerging from the major European equipment suppliers. The company also actively pursues information on its smaller European competitors, who often are open to licensing agreements for their process innovations.

Interfirm Collaborations

Collaboration with other firms has become many companies' preferred mechanism for gaining global access to technological developments. These collaborations may be formal joint ventures, interfirm agreements, participation in international consortia, or a variety of other possibilities. For instance, joint ventures between U.S. and Japanese manufacturers in the automobile and steel industries have been motivated primarily by U.S. desire to gain firsthand experience with the Japanese production system, to gain access to new product and process technology, and to learn effective implementation practices; Japanese producers have sought to gain U.S. production experience and to overcome U.S. trade barriers. In autos, joint ventures such as those of General Motors-Toyota, Chrysler-Mitsubishi, and Ford-Mazda have followed this general pattern. In the domestic steel industry, a number of joint ventures (see Table 1) have given U.S. producers advanced technology, as well as the financial backing for required investments, while giving Japanese producers broader access to the U.S. market and a production base to supply the American plants of Japanese auto producers.[13]

In contrast to decisions made to gain access to foreign technology, U.S. firms that have technological advantages often use them to increase their manufacturing and sales operations abroad. Joint ventures in the aircraft industry, for example, typically are motivated by the U.S. firm's desire for greater market access and the foreign firm's desire for technology.[14] In microelectronics, small U.S. manufacturers have entered agreements

---

[13] When Kawasaki Steel purchased 40 percent of Armco Steel in 1989, Robert Boni, chairman of Armco, stated, "The Japanese bring a technical excellence to the table that attracted us." See Jonathan Hicks, "Foreign Owners Are Shaking Up the Competition," *New York Times*, May 28, 1989, p. F9.

[14] For a full discussion of international collaborations in the commercial aircraft industry, see

TABLE 1 Major Japanese-American Joint Ventures in Steel

U.S. Company	Japanese Company	Venture
National Intergroup	NKK Corporation	Integrated producer
National Steel*	Marubeni Corporation	Slitting operation
Inland Steel	Nippon Steel	Cold-rolling mill
LTV Corporation	Sumitomo Metal	Electrogalvanizing
Baker Hughes, Inc.	Sumitomo Metal	Steel pipe
Wheeling-Pittsburgh	Nisshin Steel	Coating line
Armco, Inc.	C. Itoh	Steel processing
Steel Technologies	Mitsui	Service center

*National Steel Corp. is a joint venture between National Intergroup, Inc. and NKK Corp.

SOURCE: *Wall Street Journal*, November 18, 1988.

with large foreign firms that, in their basic form, trade U.S. design tools and circuit libraries for manufacturing capacity and distribution support abroad. More complex forms of such agreements have traded design tools and designs in certain product categories for similar tools in other products and for manufacturing expertise.

Although such agreements are reached between small and large U.S. firms, a small firm often finds foreign companies more receptive. For instance, managers at one small U.S. microelectronics firm told the committee that they had found a complex agreement with Hitachi attractive because the Japanese firm promised better cooperation than did potential U.S. partners. Similarly, U.S. aerospace suppliers and subcontractors were reluctant during the 1970s to enter risk-sharing agreements with the Boeing Company for the development of new commercial aircraft. This left Boeing little choice but to seek foreign partners. In some cases, such agreements have included capital investments by the foreigner in the U.S. firm. With or without equity investments, the emergence of technology-sharing agreements between small entrepreneurial U.S. firms and large foreign corporations creates an effective mechanism for transferring U.S. technology to foreign competitors.

Dispersion of R&D Facilities

A final example of mechanisms used to gain global access to technology

---

David C. Mowery, *Alliance Politics and Economics: Multinational Joint Ventures in Commercial Aircraft*, Cambridge, Mass., Ballinger Publishing Co., 1987.

is many large companies' maintenance of global R&D facilities in key countries worldwide. According to the National Science Foundation, spending on research overseas by U.S. companies increased 33 percent between 1986 and 1988 compared to only 6 percent at home. U.S. microelectronics firms are among the most aggressive in establishing research and engineering capabilities abroad. The motivations vary. In some product categories, such as application-specific integrated circuits, customers' demands for product customization and speedy delivery have led to the establishment of design centers in foreign countries. In other cases the availability of high-quality, low-cost foreign engineers, often trained in the United States, has been the major consideration in establishing R&D facilities abroad. In contrast, Japanese electronics, automobile, and steel firms have begun to establish research facilities in this country only recently, but the trend is apparent. Many European chemical, pharmaceutical, consumer electronics, and electrical equipment firms also have R&D labs in the United States.[15]

Firms without the resources to establish wholly-owned research facilities gain access to research results through guest researchers or students at universities abroad, analysis of patent applications, licensing of new developments, and purchasing of R&D from foreign firms. For instance, major Japanese corporations have established well-managed, comprehensive programs for monitoring current academic research, particularly in the United States, where the system of university-based research is the world's most open. U.S. manufacturers, in contrast, have been slow to build effective networks for gathering timely information on global technological developments.

CHANGES IN COST PRIORITIES

Another factor central to the internationalization process is the major changes that have emerged in the understanding of and priority given to different components of manufacturing costs. Though cost structures in different industries vary widely, an important shift is occurring in manufacturers' understanding of the role of direct and indirect labor, sources and allocation of overhead costs, the importance of capital costs, the hidden costs of poor product design, poor workmanship, and scrap and rework

---

[15] U.S. affiliates of foreign manufacturing firms spent almost $5 billion on R&D in the United States in 1986. By industry the top spenders were industrial chemicals, electrical and electronic equipment, drugs, nonelectrical machinery, primary and fabricated metals, and transportation equipment. See Bureau of Economic Analysis, *Foreign Direct Investment in the United States: Operations of U.S. Affiliates of Foreign Companies*, Washington, D.C., U.S. Department of Commerce, 1988.

(the costs of quality), and the importance of a support infrastructure in achieving production goals.[16]

In the 1960s and 1970s much of the offshore movement of U.S. manufacturing was driven by a strategy based on low labor costs. U.S. managers reasoned that the low prices of many imported products, from automobiles to consumer electronics, were attributable to cheap labor; the only way to compete was to move production offshore to match or beat those labor costs. Over time, however, the importance of labor costs in international investment and production decisions has been diminished by a number of factors:

- Advances in technology and a long history of squeezing labor out of production have reduced direct labor content in most manufacturing industries to 15 percent or less of production costs; in high-technology industries it seldom exceeds 10 percent and increasingly is under 5 percent.[17] Even wide variations in direct labor costs have relatively insignificant effects on total costs.

- Many companies have discovered that a strategy based on low labor costs is difficult to implement in the long term. Low-wage locations are equally open to competitors' investments, relative wage levels between locations can change abruptly with changes in exchange rates and shifts in demand, and the extra capital and logistical costs of shifting operations to alternative sites to maintain low wages (so-called "island hopping") can easily consume the resulting wage savings.

- In most manufacturing industries, labor costs have been outweighed by market access, quality control, timely delivery, and responsiveness to customers as determinants of global competitiveness. Competitive manufacturers, adopting manufacturing systems for total quality control, have found that siting plants in areas with the skills and capabilities needed to control **total** costs and quality is more important than siting them to minimize labor costs. These considerations explain the predominance of developed economies in U.S. manufacturing investment abroad (see Figure 7).

Direct labor still receives more attention than it deserves because of accounting systems that continue to allocate overhead costs on the basis of labor load. Still, realizations of the true significance of labor costs are beginning to have important effects on management decisions. Particularly for firms pursuing total quality, international differences in the costs of assuring quality in production are the critical factor in investment decisions.

---

[16] The Manufacturing Studies Board is currently engaged in a major study of international differences in the cost factors driving manufacturers' global production and investment decisions.

[17] Clark, Hayes, and Wheelwright, *Dynamic Manufacturing*, p. 137.

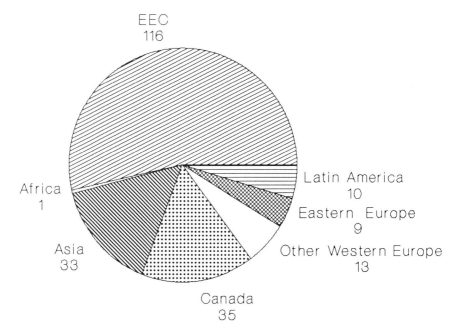

FIGURE 7 U.S. manufacturing investment abroad, 1989: acquisitions, joint ventures, new plants.

The pursuit of total quality in different manufacturing processes reveals a wealth of information about where costs are created and the resources and capabilities needed to control them. The ability to minimize inventory, to eliminate scrap and rework, to integrate functions, and to shorten time to market depends on trained personnel, a strong supplier base, a streamlined organization, and flexibility. Therefore, international site selections might hinge on the cooperation, flexibility, and trainability of the labor force rather than on its cost. For firms using just-in-time systems, the availability of a responsive supplier base or the ability to create one quickly is another important consideration.

The lessons learned about a labor cost strategy and a total system approach to cost control have created different considerations in production location decisions, but direct labor remains important to production costs in some product lines. Managers from several industries told the committee that, as a general rule, if direct labor costs exceed 10 percent of total costs, production can be cheaper in low-wage locations. Assembly of electronic devices is one example. Many of the large companies in this industry use sophisticated models to determine the relative advantages of using automated assembly technologies (when available) or low-cost labor at various wage rates; proximity to customers and suppliers is also considered as well as various scenarios for overhead costs. The results vary by product. For instance, low-cost labor may be preferred to automation for products with very short model lives to minimize fixed asset exposure. When low-cost labor offers a clear advantage, sites in Thailand, Malaysia, and the Philippines are attracting more U.S. and Japanese investment because labor costs in Korea, Taiwan, and Singapore have increased rapidly in recent years.

One kind of labor whose costs have begun to receive more attention, at least in knowledge-based industries, is highly skilled technical labor for support services. For instance, Texas Instruments has joined AT&T and other U.S. firms in establishing software engineering facilities in India to take advantage of well-trained, low-cost computer programmers. Other firms are finding that the engineering labor force in Korea, Singapore, and other NICs compares favorably to U.S. engineers but at a fraction of the cost. However, managing high-value offshore activities effectively remains a challenge, particularly product and process engineering activities that often are critical to competitive success. On one hand, the short-term cost advantages of moving product/process design and engineering responsibilities to low-cost locations, where firms often have factories, are evident. On the other hand, the broad infrastructure necessary to maximize engineering's value added may not exist in low-wage locations, and the ability to integrate offshore engineering into the firm's total manufacturing system may be constrained by distance and cultural differences. Consequently, the long-term effects remain unclear.

In some capital-intensive industries, such as semiconductors and chemicals, the cost of capital is the major cost consideration in investment decisions.[18] Because labor needs are minimal, supplies are shipped globally, and transportation costs are low, the key factors in site selection are availability of low-cost capital and the presence of the appropriate technical and logistical infrastructure. In such capital-intensive industries, government incentives to reduce capital costs using mechanisms such as interest rate subsidies, tax holidays, and cost sharing on plant and equipment can be particularly effective in swaying corporate location decisions. Though the advantages of such incentives to individual companies are undisputed, it

---

[18] A number of recent studies have addressed international differences in the cost of capital as a source of U.S. disadvantage in manufacturing competitiveness. For instance, see Robert McCauley and Steven Zimmer, "Explaining International Differences in the Cost of Capital," *Federal Reserve Bank of New York Quarterly Review*, vol. 14, no. 2, Summer 1989, pp. 7-28. The committee's intent in this brief discussion is not to resolve the complex issue of whether U.S. capital costs are higher than those of other nations but to note the importance of access to low-cost capital as a driver in the internationalization process.

remains an open question whether these government programs represent an efficient use of public resources.[19]

POLITICAL AND ECONOMIC FACTORS

In addition to changes in markets, technology sources, and cost structures, a variety of political and economic factors have changed the global manufacturing environment. The range of factors extends from exchange rate volatility to specific government-to-government agreements both to open markets and to manage trade. This aspect of the internationalization process is far too extensive to permit an exhaustive review here, but a few examples will illustrate the importance of such factors to manufacturers' international decisions.

Exchange Rate Fluctuations

The 1971 breakdown of the Bretton Woods system of fixed exchange rates introduced a new source of risk to international trade and investment. Variations in exchange rates under the current floating rate regime have been large relative to typical profit margins; changes of 1 percent in a day and 20 percent in a year are not uncommon. Many economists argue that this volatility is more attributable to the portfolio preferences of investors than simply to adjustments needed to balance trade flows. Because shifts in exchange rates are unpredictable, manufacturers are faced with high uncertainty in the costs and returns of many international operations.

A wide array of financial mechanisms are available for hedging against currency fluctuations, but they are most effective for short-term variations. Manufacturers have responded to long-term currency shifts by diversifying production geographically, making exchange rate variations an integral factor in the internationalization process.[20]

For instance, the dramatic cost reductions achieved by major Japanese exporters in recent years were necessitated by the tremendous rise in the value of the yen after 1985-1986. Such cost reductions have not

---

[19] The question of the costs and benefits of government investment incentives typically is addressed in the context of local public finance debates. Incentives are typically provided by state, regional, or local governments, both in the United States and abroad, though some national governments, such as Ireland, have made incentives a matter of national policy. For a discussion of many of the issues involved in the level and form of government investment subsidies, see, for example, R. Scott Fosler, "State Economic Policy: An Assessment," *Business in the Contemporary World*, Summer 1989, pp. 86-97.

[20] These arguments are made more extensively in Ronald I. McKinnon, "Monetary and Exchange Rate Policies for International Financial Stability: A Proposal," *Journal of Economic Perspectives*, vol. 2, no. 1, Winter 1988, pp. 83-103.

only been applied to domestic manufacturing operations but also have led to more offshore production and purchasing by Japanese corporations. The effect has been to accelerate the internationalization of the Japanese manufacturing base.[21] U.S. companies pursued similar responses to the strong dollar of the early 1980s. Another response is for companies to establish production facilities in local markets to minimize their dependency on exports to that market and, therefore, to reduce the effects of currency swings on product prices and market share. This reasoning has been an important motivation for the recent increases in Japanese manufacturing investment in the United States and Europe. Though reducing exposure to currency swings may not be the primary driver in decisions to locate production facilities or to buy components in specific countries, it can speed such decisions and make an international production strategy both unavoidable and competitively sound.

Europe 1992

The multigovernment initiative under way to create a unified European market by 1992 is dramatically changing corporate plans and strategies in Europe. The prospect of a single market of 320 million consumers is already spurring new investments, plant rationalization, and strategic alliances that would be unlikely otherwise. For many manufacturers in Europe, 1992 offers opportunities to consolidate small plants built to serve individual national markets into European-scale facilities. For firms that have had no production facilities in the European Community, the fear that 1992 may bring greater trade protection in Europe is prompting reevaluation of the viability of export strategies and boosting investment plans. Europe 1992, more accurately described as a process than as a specific event, is stimulating changes in strategies among manufacturers worldwide and will change the role of Europe in an increasingly interdependent global economy.[22]

U.S.-Canada Free Trade Agreement

The U.S.-Canada Free Trade Agreement of 1988 is sparking similar changes in the North American market. Both U.S. and Canadian companies have announced rationalizations in output and steps to determine

---

[21] Japan's offshore production is expected to be 8.7 percent of total Japanese output in 1990 compared to 3.5 percent in 1986. See "Remaking Japan," *Business Week*, July 13, 1987, p. 54, and Eileen M. Doherty, *Japan's Foreign Direct Investment in Developing Countries*, Japan Economic Institute, August 11, 1989.

[22] For an instructive survey of the 1992 process and its implications for business, see Nicholas Colchester, "A Survey of Europe's Internal Market," *The Economist*, July 8, 1989.

which plants will serve what market. The resulting integration of the two economies will further complicate the already difficult problem of accounting for trade between the two nations.

Trade Protection

The General Agreement on Tariffs and Trade (GATT) has been very successful in achieving major multilateral reductions in tariffs since World War II. In terms of tariffs alone, the level of trade protection has fallen dramatically, resulting in the virtual elimination of tariffs for most industrial goods traded among developed countries.[23] The predictable result has been explosive growth in world trade. With GATT deterring tariff increases or additions, however, nontariff barriers to trade have become a widely used alternative for many national governments. Nontariff barriers take a variety of forms, resulting in a range of responses by manufacturers.

Two forms of nontariff barriers—voluntary export restraints and trigger price mechanisms—have been used by the U.S. government in recent years. The use of voluntary export restraints (VER) on Japanese automakers effectively imposed a quota on U.S. imports of Japanese cars. The Japanese response has been (a) to upgrade the mix of cars exported to this country to maximize revenue per unit sold, and (b) to invest in U.S. production capacity to build market share unconstrained by VER agreements. President Reagan initiated 5-year VER agreements on machine tools with Japan and Taiwan in May 1986. In this case the foreign response has been to purchase existing domestic capacity, to build new manufacturing facilities in this country, to license domestic builders, and to create joint ventures with American machine tool builders.[24] The government has used trigger price mechanisms—establishing a minimum price for sales in the U.S. market—in both semiconductors and steel as a way to eliminate dumping by foreign manufacturers. The result has been excess profits for low-cost foreign producers, allowing greater investment to reduce production cost, to advance the level of product and process technology, and to build manufacturing capacity in the United States.

Foreign governments have used similar nontariff barriers with similar effects on the strategies of foreign manufacturers. Europe is a good example. Japanese companies have dramatically increased their manufacturing

---

[23] A current GATT tariff study gives these average tariff levels (in percent of value of imports) for selected countries: Canada—6.9, United States—4.2, the European Economic Community—4, Sweden—3.8, Japan—2.7, and Switzerland—2.2.

[24] For additional information on the internationalization of the U.S. machine tool industry, see U.S. Department of Commerce, *U.S. Industrial Outlook*, recent years, and Nicholas S. Vonortas, *The Changing Economic Context: Strategic Alliances Among Multinationals*, Troy, NY, Rensselaer Polytechnic Institute, 1989.

investments in Europe—for example, in automobiles (Nissan, Toyota, and Honda in Great Britain), semiconductors (Fujitsu in England), and consumer electronics (Matsushita, Toshiba, Hitachi, Sony, and Canon, among others, in Germany, France, Spain, and England). U.S. companies have also established manufacturing facilities to breach trade barriers. For example, both Texas Instruments and Intel plan to build new semiconductor fabrication facilities in Europe (Italy and Ireland, respectively) in response to increases in the amount of semiconductor processing that must be done in Europe to avoid import restrictions.[25]

Local content requirements, as illustrated in the European semiconductor example, are a particularly prevalent form of nontariff barrier. Such requirements historically have been used to support local development of key industries. Consequently, they have been an important factor in foreign direct investment decisions and a potent driver of internationalization. In some cases, local content requirements may preclude penetration of a specific market. The firm may deem the sales potential in the market too small to justify a manufacturing investment or may view the local supplier base as inadequate to fulfill quality specifications. In other cases, market presence may be essential to tap sales potential or to gain access to specific resources, but the local content rules force decisions on the amount and form of manufacturing investment that are suboptimal from the firm's perspective. The rules introduce constraints on management flexibility, dictate the type and scale of technology used in certain locations, and preclude desired plant and product rationalizations. Finally, local content requirements may interact with other trade barriers to complicate the ability to export from a given plant. Examples are the recent disputes in Europe over the nationality of cars produced by Nissan in England and photocopiers produced by Canon in California. The value added locally to these products is deemed insufficient to qualify them as English or American; the final products therefore would be deemed Japanese and subject to existing trade barriers. Such disputes may proliferate as governments find it increasingly difficult to define a local product unambiguously as companies integrate production of components, subassemblies, and final goods into global manufacturing systems.

Nontariff barriers take a range of forms in addition to those described above. They include public sector policies such as government procurement, subsidies, standards, regulations, and patent policy. They also include private sector moves such as supplier qualification, distribution constraints, collusion, and oligopoly. Each type of barrier tends to have characteristic

---

[25] "Intel Will Make Chips in Ireland," *Semiconductor Industry and Business Survey*, October 9, 1989, and Stuart Auerbach, "Europe 1992: Land of Opportunity Beckons," *Washington Post*, March 20, 1989, p. A 1.

effects on firms' international production and marketing strategies, particularly in high-technology industries.[26] For instance, many foreign firms use joint ventures with local firms to overcome nontariff barriers in Japan. As these nontariff barriers affect manufacturers striving to compete on a global basis, the emphasis on them in trade negotiations is likely to continue to increase, resulting in further blurring of trade policy and domestic policy and continued disputes about the proper role and mechanisms for each.

CONCLUSION

Markets, technology, costs, and politics have been the major forces for change in the international manufacturing environment in the past 10 to 15 years. Market access has become the dominant driver of international investment decisions, but other factors retain varying influence in specific situations. Low-wage locations are still advantageous for manufacturing products with high labor content. Companies that base their competitiveness on technological leadership give priority to access to technology. Other companies may focus on global proliferation of production capacity to speed response to customers and to tailor products to local demand. Firms with high capital requirements, in contrast, need to centralize production. The relative weight given to each factor varies by company and product and over time. No two companies face the same international challenge.

Managers at several U.S. multinational corporations told the committee that their primary challenge in coming years is integrating their existing global operations to perform as a single system rather than as islands of manufacturing and technological capabilities. These firms have extensive international networks of marketing and production facilities, based primarily on a multidomestic model of international business. In this model, foreign operations tend to duplicate those of the parent and are given substantial autonomy over production, distribution, product development, and research. The resulting global production base, rather than U.S. exports, has been the vehicle for U.S. penetration of global markets. Few companies have progressed far in coordinating this extensive infrastructure on a global basis, but the emergence of advanced communication and manufacturing technologies has made global integration possible.

The challenge of global integration is magnified by the still poorly defined nature of prospective global organizational systems. In the computer and information industry, for instance, rapid escalations in the costs of

---

[26] For a full discussion, see Michael F. Oppenheimer and Donna M. Tuths, *Nontariff Barriers: The Effects on Corporate Strategy in High-Technology Sectors*, Boulder, Colo., Westview Press, 1987.

R&D, capital equipment, software, and customer support are forcing companies to strengthen their global market presence and to integrate global operations to eliminate redundancy. At the same time, the global diffusion of technological and scientific expertise in the industry calls for ready access to sources of innovation around the world. This need tends to restrict the ability to rationalize global operations to achieve efficiencies. Balancing these two pressures appropriately has become a fundamental challenge to the industry. Managers in most other industries face similar challenges in creating integrated global manufacturing systems.

For emerging multinational companies, the challenge is not to integrate but to expand global presence. Managers must consider the types of marketing, distribution, and manufacturing presence necessary to build foreign market share; the advantages of various locations; the possibilities for partnerships with domestic companies (and in some cases the necessity of partnerships); the source and level of control of foreign production; and the likely impact on the existing production base. Careful attention to these issues, with decisions based on accurate information, can avoid investments that fail to perform as intended or alliances that give up more than they gain.

Small and mid-sized U.S. companies also face the challenges of building an international infrastructure. Often they are suppliers to large corporations striving to penetrate global markets, and in some cases, such as auto parts, they are subject to direct competition from new foreign investment in this country. The growth of international competition in the markets served by small firms has set new standards for their production operations and put a premium on staying abreast of technological developments worldwide and attacking foreign markets. Most small firms lack the resources to spread their production base to foreign markets, but options are available. For instance, supplier firms can forge closer links with customers to take advantage of larger firms' resources and international experience whenever possible. Direct experience abroad can be gained by expanding exports.

Opportunities for foreign sales are often underestimated. A 1987 report from the American Business Conference demonstrated that small to mid-sized firms—the bulk of the U.S. manufacturing base—can succeed in building international sales, given management commitment.[27] For the firms studied in this report, the first step was to build foreign distribution capabilities, leading to overseas production once sales reached a critical mass. The required initial investments were fairly small, the first target markets were often English-speaking ones to minimize cultural risks, and success was quick—the companies studied increased their foreign sales nearly 20 percent annually in the 1980s, and most showed profits on foreign

---

[27] American Business Conference, Inc., *Winning in the World Market*, Washington, D.C., 1987.

sales within the first year. The success of these companies clearly indicates the potential for greater foreign sales by U.S. manufacturers, given the required commitment. For firms without their own sales networks, export trading companies have been formed to provide marketing and distribution services. Learning what is necessary to serve foreign customers through exports is probably the most important step for small firms wanting to benefit from the internationalization process.

Japanese manufacturers tend to face a somewhat different challenge than U.S. firms. With only a few exceptions, the Japanese have relied on an export strategy to achieve broad penetration of world markets. Though their marketing and distribution infrastructure is well established, they now face the need to create a global manufacturing infrastructure. Changes in the value of the yen, improving capabilities of other Asian nations, and fear of increased protectionism in world markets are spurring a dramatic increase in Japanese offshore production. By far the preferred mechanism has been wholly-owned greenfield investment, with North America the favorite site, but joint ventures with local producers have been common in industries such as automobiles, steel, and machine tools. Many of these facilities continue to be assembly operations that receive components from Japan (standard practice for new plants by multinationals of all nationalities), but several firms have said they intend to increase local value added in the next few years. Honda, for example, plans to introduce to the U.S. market a new automobile designed, engineered, and manufactured entirely in the United States.

European manufacturers face the same kinds of issues. Major European multinationals, like similar U.S. firms, have extensive international marketing and manufacturing capabilities; their main challenge is coordinating these global operations. To a greater degree than their U.S. counterparts, however, European multinationals have tended to retain management control and strategic direction for global subsidiaries at their European headquarters, creating a somewhat different basis for building a globally-integrated manufacturing system. European firms that are multinational only within the European market face the challenge of expanding their marketing and manufacturing presence in other world markets.

To an increasing extent, effective responses to the causes of internationalization will push manufacturers from each region toward similar objectives. The need for a global marketing scope is well understood, but the importance of a global manufacturing scope is only beginning to be realized and the means of achieving it are still not well understood. The differences in starting points among various companies and nationalities will sometimes constrain the tools available and color the challenges to be overcome. Because they have the most extensive global marketing and manufacturing infrastructure,[28] U.S. companies are potentially in an advantageous position. Unfortunately, too few U.S. firms recognize the value of their existing infrastructure or the steps needed to strengthen it.

---

[28] According to the Federal Reserve Board, all U.S. assets abroad currently continue to exceed foreign assets in the United States by a wide margin if assessed at market value, $785 billion versus $466 billion.

3
National Economic and Political Implications

The initiatives being taken by manufacturers to succeed in the global economy are creating networks of assets, technology sources, markets, suppliers, and partners that transcend national borders. Building such networks, or at least participating in them, is becoming a critical factor in competitive success, not only for multinational corporations but also for smaller companies being challenged in domestic markets. Although it is difficult to predict all the consequences for specific companies or the national manufacturing sector, a few trends seem clear.

INWARD INVESTMENT AND FOREIGN OWNERSHIP

The importance of foreign investment flowing into the U.S. manufacturing sector will continue to grow. Although foreigners now own roughly 12 percent of total U.S. manufacturing assets, the percentage can be much higher in specific industries. For instance, foreign firms control about one-third of U.S. production in the chemical industry and about 40 percent in the tire industry.[1] In the most visible case, Japanese production of automobiles in U.S. factories is forecast to account for 12 to 15 percent of domestic production by 1991.[2]

---

[1] Graham and Krugman, *Foreign Direct Investment in the United States*, p. 33.

[2] Jonathan P. Hicks, "Foreign Owners Are Shaking Up the Competition," *New York Times*, May 28, 1989, p. F9.

Tangible reasons to encourage inward flows of foreign investment: Honda has been steadily increasing the American content of cars built at its Marysville, Ohio, plant, requiring demanding quality and delivery improvements by U.S. suppliers to the benefit of their total business. Stamped panels in production inside Honda's Marysville Auto Plant.

These investment inflows have become a critical part of the U.S. manufacturing sector. They provide capital to help maintain the size of the manufacturing base, introduce competition to reinvigorate domestic firms, and in many cases displace imports or even produce exports. Such investment flows also bring closer contact with foreign management practices, investment procedures and priorities, and organizational relationships that benefit domestic firms. For instance, Honda has been steadily increasing the U.S. content of the cars built at its Marysville, Ohio, plant, requiring demanding improvements in quality and delivery by U.S. suppliers to the benefit of their total business. Inland Steel gained additional customers as a result of working with Honda to improve the quality of its coated steel.[3] Such indirect benefits, combined with capital needed for manufacturing investment, provide tangible reasons to encourage inward flows of foreign investment.

Unfortunately, foreign direct investment has evoked considerable criticism for two major reasons. First, there is the fear that the United States is losing control of its productive assets, endangering national security. A

---

[3] Doron P. Levin, "Honda Blurs Line Between American and Foreign," *New York Times*, March 14, 1990, pp. 1, D8.

number of concerns surface in this context. They range from the role of foreign affiliates during wartime, to the potential loss of foreign component supplies, to the loss of control of critical technologies, yet the true degree of threat in these areas has not been systematically analyzed or confirmed. On one hand, political intervention, justified on the premise of controlling critical technologies, has squelched a few foreign acquisitions, such as Fujitsu's attempted purchase of Fairchild Industries in 1988. On the other hand, little effort has been made to identify critical technologies outside the defense context, to determine the status of U.S. technology relative to foreign technology, or to identify the range of technologies in which foreign investment should be encouraged to strengthen national capabilities.

The second major criticism of foreign investment is that foreign affiliates in the United States are primarily assemblers of imported components, providing low-wage jobs and little value added. The evidence does not support this concern. Compared to U.S. firms in the same industries, foreign affiliates pay comparable wages, conduct similar levels of R&D, and add comparable value.[4] In fact, the majority of foreign investment goes into the acquisition of existing firms, often resulting in an increase in capital spending for facilities' improvement.

The predominance of these concerns in the national debate does a national disservice. Instead of decrying unsubstantiated negative effects of inward investment—echoing European concerns about U.S. investment in the 1960s—national attention should be focused on creating a favorable environment for investments in high-value manufacturing activities, regardless of the source. Product and process design and implementation, engineering, and R&D provide the basis for innovation and continued participation in emerging markets. These types of high-value activities, combined with production of high-value products, must be pursued in the United States to maintain vigorous economic growth.

As corporations continue to confront the pressures of global competition, managers will make decisions intended to maximize the competitiveness of their firms. The competitive environment will not allow U.S. firms the luxury of siting manufacturing activities in this country for patriotic reasons, and the size of the U.S. market is no guarantee that foreign firms will undertake high-value activities here. Managers will locate these high-value activities where the intellectual expertise and business infrastructure exist to perform them most effectively and profitably. This basic fact must govern the decisions of policymakers. Policy initiatives must focus on strengthening the national infrastructure needed to support high-value activities and nurture new business development, rather than addressing

---

[4] Graham and Krugman, *Foreign Direct Investment in the United States*, pp. 48-54.

unsupported and misplaced concerns about ownership and control of the U.S. manufacturing base.

TECHNOLOGY FLOWS

A second clear consequence of the internationalization process is that it has become impossible to control the flow of technology across national borders. This development is unsettling from the point of view of a nation like the United States that bases its international competitiveness on advanced technology. The prevalent tone of national debate has emphasized the need to control the flow of technology abroad using export controls or exclusion of foreign firms from R&D consortia. For instance, the Microelectronics and Computer Technology Corporation (MCC), a private research consortium based in Austin, Texas, limits membership to companies with more than 50 percent U.S. or Canadian ownership. The issue is more complex, however, when the consortium receives government funding and blocks foreign membership, which is the case for the National Center for Manufacturing Sciences (NCMS) and for Sematech.[5] Government funding makes foreign membership an issue of national policy, and forbidding it can prompt retaliation by foreign governments. For instance, the Joint European Submicron Silicon (JESSI) research effort is closed to U.S. companies.

Such attempts to limit foreign access to U.S. civilian technology ignore the realities of both the U.S. and international economies. The relationships among companies, the presence of companies in every major market, the dispersion of technology-creating activities across countries, and the free flow of goods have made technology an international resource. Furthermore, the open education system and the relationship between industrial and academic research in this country give foreigners easy access to U.S. basic research. In fact, the predominance of foreigners in U.S. graduate schools of science and engineering makes such research strongly dependent on them (see Figure 8). Instead of trying to restrict foreign access to U.S. technology—an impossibility—U.S. energy should focus on facilitating the flow of foreign technology into the United States by ensuring comparable access to foreign research and conducting effective intelligence on foreign technological developments, and, most importantly, on speeding the process of turning technology into commercial products.

---

[5] For membership purposes, NCMS requires companies to be "U.S.-based with a substantial portion of their research, development, and manufacturing occurring within U.S. boundaries. In addition, U.S. citizens must hold majority ownership and control...."

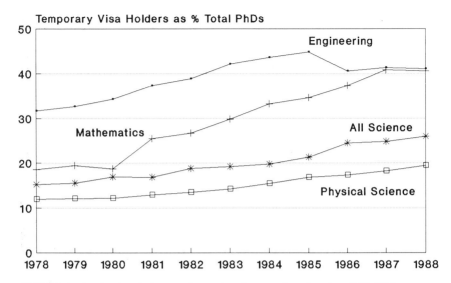

FIGURE 8 Ph.D.s awarded to foreigners in science and engineering, 1978-1988.

DOMESTIC VERSUS INTERNATIONAL POLICY

Internationalization is blurring the distinction between domestic and international economic and technology policy. In a growing number of cases, policies developed with little consideration for their effects on trade have become targets for trade negotiations. Japan, for instance, has the lowest tariffs of any industrial country, but her domestic policies and customs determine the openness of the market to foreign products. A good example is the recent negotiations on Motorola's access to the cellular phone market in Tokyo and Osaka, which hinged on the allocation of radio frequency rights in those cities rather than on import restrictions on foreign cellular phones. In the United States the linkage has become particularly apparent in technology policy.[6] The recent debate on high-definition television (HDTV) provides an example. The Federal Communication Commission's decision to require HDTV transmissions that will be compatible with existing receivers effectively precluded use of emerging Japanese and European transmission standards. The decision will certainly slow, though not eliminate, Japanese and European penetration of the U.S. market and will require additional development of U.S.-compatible equipment.

The potential for conflict between domestic and international policy is

---

[6] For a discussion of many of the issues involved in technology policy and internationalization, see David C. Mowery and Nathan Rosenberg, "New Developments in U.S. Technology Policy: Implications for Competitiveness and International Trade Policy," *California Management Review*, Fall 1989, pp. 107-124.

particularly apparent in regulatory policy. Virtually every type of regulation, from control of smokestack emissions to mandatory worker benefits to antitrust, has some impact on the ability of U.S. manufacturers to compete globally. When the U.S. economy was fairly self-sufficient, regulatory policy could be imposed with the knowledge that all major producers would be treated equally. Although foreign firms manufacturing in the United States are subject to the same regulatory regime as domestic firms, internationalization has increased the number of foreign competitors operating under different and potentially less costly regulatory conditions and created opportunities for domestic firms to produce abroad to avoid some types of U.S. regulations. Such considerations should play a greater role in setting U.S. regulatory policy to avoid making this country a high-cost manufacturing location unnecessarily.

INADEQUATE INFORMATION

Another consequence of internationalization is that public and private statistical data often do not capture the breadth of information needed to formulate effective policy. Existing government statistics have become inadequate to reflect the complex relationships among the United States as a country, U.S. corporations, and the rest of the world. For instance, current government reporting requirements, data gathering methods, and statistical interpretations are not designed to take account of the growing percentage of trade conducted on an intrafirm basis. When a firm ships U.S.-made components abroad for final assembly, reimports some of the final product, and exports the rest to third markets, the transaction is far more complicated and represents far more export value to the U.S. firm than is shown in standard trade statistics.[7] Relatively minor adjustments to an individual firm's production system can create dramatic changes in the accounting of national imports and exports. Such inadequacies in the existing trade statistics indicate that the trade balance is flawed as the measure of national competitiveness or the definer of needed policy measures. A more accurate picture of national competitiveness could be gleaned from data on factors such as global market shares of U.S. companies, nature and location of manufacturing activities, and sources of materiel supply.[8]

---

[7] William Finan, "Globalization Is Skewing the Trade Statistics," *The International Economy*, January/February 1988, p. 132, and Paul Blustein, "Critics Say U.S. Economic Picture Is Blurred by Reliance on Bad Data," *Wall Street Journal*, January 10, 1986, p. B1.

[8] For an overview of the factors influencing the accuracy of national statistics, see Office of Technology Assessment, *Statistical Needs for a Changing U.S. Economy*, Washington, D.C., U.S. Government Printing Office, 1989.

CONCLUSION

In an earlier era, when the U.S. economy was largely self-sufficient, domestic policy could be set with little consideration of its international implications. Both public and private policymakers could make decisions with the confidence that the primary competitors in the U.S. market were all domestic, that the sources of new technology were domestic, and that the only information needed to set sound policy was available nationally. Similarly, foreign policy was shaped by this confidence in U.S. superiority, emphasizing free trade and open access to and national treatment for U.S. direct investment abroad. For many years the United States was relatively immune from the reciprocal consequences of these policies.

Internationalization has changed the context for U.S. policy. Historic policies have not only expanded world trade, spurred the growth of U.S. multinational corporations, and advanced global economic development but also have increased this country's integration into the global economy. Awareness of the domestic policies of major trading partners, foreign technology developments, and the global flow of technologies has become essential to effective policymaking. The need for additional information that accurately portrays the position of U.S. manufacturers and the U.S. economy has increased at the same time that characterizing a company as U.S. or foreign has become more difficult. These factors have heightened the need to define national interests clearly with the recognition that the United States has much to gain from foreign manufacturers and much to lose from restrictions on foreign competition.

The fundamental premises of U.S. policy—free flows of goods, investment, technology, and scientific knowledge–remain valid. Recognizing the benefits and the necessity of inward flows of investment, technology, and goods has become more important than propounding the virtues of outward flows.

4
Keys to Success

Internationalization has created unprecedented competition for U.S. manufacturers. Success requires that quality, cost, service, and value be truly world class, regardless of the size of the firm or type of industry. A number of factors can be identified that have become essential to the competitiveness of U.S. manufacturing companies and the nation's manufacturing sector. Though specific steps needed to maintain global competitiveness will vary by company, the following themes can be applied across the board as the foundation of corporate and national success in the internationalization process.

SOURCES OF CORPORATE SUCCESS

Internationalization does not change the metrics of corporate success. The fundamental importance of profitability in the U.S. system is not likely to change, but corporations will have to recognize the need to place less emphasis on short-term profits and more emphasis on participating in international markets in positioning themselves for long-term profitability. In this light three other characteristics can be linked to corporate success in the international environment. First, the firm must maintain control of corporate destiny by retaining control of the core capabilities necessary to support a strong long-term presence in its chosen markets and technologies. Second, the firm must be able to benefit from changes in technology, industrial structure, and other significant events. Third, the firm must nurture the stakeholders in the business, including employees, suppliers, customers, communities, and shareholders. A manufacturer that can achieve these objectives and maintain long-term profitability will benefit from the internationalization process. The challenge is to use internationalization as a tool to strengthen the firm's abilities in these areas.

Corporate Imperatives

Drawing on its research, observations, and experience in international manufacturing, the committee identified several factors necessary for long-term success in the emerging global marketplace. As usual, specific implementation steps vary, but the concepts apply universally.

1. Develop necessary managerial capabilities

Fostering international experience should be made an explicit part of management development planning. Despite the impact of internationalization on the broad base of U.S. manufacturing, too few manufacturers have managers with significant international experience or understanding of the international marketplace. A few multinational companies emphasize foreign experience, but too many firms retain a culture that both discourages foreign assignments and fails to take advantage of foreign management experience. U.S. companies need to implement an explicit practice of rotating managers to a variety of foreign locations to impart a broad perspective of unique international markets. Such a practice should be applied, at a minimum, to manufacturing and marketing managers since the need to recognize customers' needs in diverse global markets and respond to them quickly and effectively will be prerequisites to corporate success.

Broader experience in international operations is not the only need in management development. Steps are also needed to improve the management of technology in U.S. corporations. Again, some firms are very good at developing, acquiring, and leveraging technological assets, but, overall, technology management remains a weakness. Managers need to do a better job of

- identifying the key technologies that determine the firm's ability to compete in specific product segments now and in the future,
- creating the capability to exploit existing expertise and update that expertise constantly,
- assessing the technological capabilities of competitors and tracking relevant technological developments from sources around the world, and
- orchestrating this entire body of knowledge to maintain a technological advantage.

2. Accumulate and exploit key competitive capabilities

Given that resources—such as funds, talent, technologies, distribution networks, and marketing information—are becoming both equal and mobile among international competitors, the only viable route to long-term success is to develop proprietary competitive capabilities that allow the firm to perform critical activities better and faster than the competition. Key competitive capabilities encompass an array of tangible and intangible strengths that build on each other and provide the ability to create innovations, to take advantage of them rapidly, and to respond to the challenges of competitors by quickly introducing products of competing cost, quality, and functionality. These strengths stem from a combination of technology, corporate organization, and collective learning abilities that are difficult to imitate because they are deeply imbedded in the firm. For example, 3M Corporation applies its expertise in substrates and coatings to create a leading position in businesses ranging from magnetic tape to masking tape. Honda considers strong capabilities in internal combustion engine technology to be critical to the global success of its businesses. Casio uses its proficiency in miniaturization across an expanding product line, including a pocket-sized photocopier.[1]

In the current manufacturing environment of rapid technology flows, high investment costs, and global competition, the ability to accumulate and strengthen key competitive capabilities is typically beyond the capacity of in-house resources. Managers must explicitly assess the capabilities the firm has today, which of them will be critical in the future, and what other capabilities need to be built or acquired to ensure long-term competitiveness. A critical question is which capabilities must be retained, practiced, and owned by the firm and which can be obtained from outside. The answer requires a holistic view of the firm's products, markets, and long-term strategy. For instance, the decision to manufacture or buy a specific part cannot be based only on cost. It also requires an assessment of the part's importance to the final product; the importance of the final product to corporate success; opportunities to gain competitive advantage through improvements in the part; and the design, engineering, and manufacturing capabilities gained as a direct result of making that part. For example, General Electric used such considerations in its decision to redesign its large refrigerator compressors and to upgrade the manufacturing systems used for their production rather than close the existing compressor line, buy from outside suppliers, both domestic and foreign, and risk losing both

---

[1] The concept of key competitive capabilities, or core competences, is more fully explored by C. K. Prahalad and Gary Hamel in "The Core Competence of the Corporation," *Harvard Business Review*, May-June 1990, pp. 79-91.

General Electric used a holistic view of the firm's products, markets, and long-term strategy in its decision to redesign refrigerator compressors to allow competitive manufacture in-house rather than source from Japan and risk losing both the design and manufacturing expertise necessary to participate in the compressor market.

the design and manufacturing expertise necessary to participate in the large refrigerator market.[2]

Strong capabilities in manufacturing-related functions will be increasingly critical to international success. Building a system of constant improvement in manufacturing is an example of a core competitive capability that provides unique advantages. Such a system must include a process of benchmarking existing manufacturing capabilities against both competitors and emerging technological developments, creating an ability to absorb new

---

[2] This decision-making process is described extensively by Ira Magaziner and Mark Patinkin in *The Silent War: Inside the Global Business Battles Shaping America's Future*, New York, Random House, 1989, pp. 67-100. Although deficiencies in the initial redesign led to compressor failures, necessitating purchases of compressors from outside suppliers while design corrections were made, the new manufacturing systems continue to perform well and General Electric has gained invaluable experience in mass producing extremely close tolerance parts. See Thomas F. O'Boyle, "GE Refrigerator Woes Illustrate the Hazards in Changing a Product," *Wall Street Journal*, May 7, 1990, pp. A1, A5.

manufacturing techniques and technologies into existing processes, and ensuring that manufacturing improvements feed through to other functions and activities in the organization. This feedthrough should occur both horizontally and vertically. Horizontally, the interaction between design and manufacturing capabilities can be a strong basis for unique competitive advantages: improvements in production processes foster improved product designs, and design improvements, demanded by customers or to match competitors, force further advances in manufacturing processes. Vertically, manufacturing capabilities at the final product level drive capabilities at all the intermediate levels from subassemblies and parts to materials, tools, and methods. The firms that are best at building key capabilities in manufacturing and integrating them horizontally and vertically throughout the firm will have a strong basis on which to build world-class competitiveness.

3. Collect and exploit global intelligence

A specific activity that U.S. manufacturers have tended to neglect is systematically gathering and using information on competitors, technological developments, and emerging product opportunities. Too many U.S. manufacturers remain provincial, focusing strictly on the domestic market in terms of market opportunities, assessments of competitors, and investment decisions. Few U.S. firms keep tabs on developments in science and technology in universities and other firms in the United States. Instead, intelligence gathering continues to have a marketing focus; efforts to gather relevant manufacturing and technological data remain haphazard and largely ineffective.

Perhaps the most significant implication of internationalization is that U.S. manufacturers must broaden their frame of reference to incorporate global competitors and opportunities. They pay little attention to foreign competitors until they face a direct threat in the United States; they make no attempt to gain access to foreign technologies, and too few of them track external technological developments. Such a cavalier attitude to competitors and technologies that will likely affect the business is unaffordable in today's international markets. U.S. manufacturers must build a systematic global intelligence network to collect coherent, strategically useful information from both domestic and foreign sources. The capacity to tap diverse sources of information, to absorb new ideas, and to integrate them into both operations and strategy has become critical to competing successfully in the international market.

Starting with an assessment of the information the firm already has or can obtain readily, managers and workers should determine

- what additional information is needed on markets, competitors, technologies, politics, and economics that will affect their business;

- what means are available for acquiring such information cost effectively; and
- what procedures, structures, and systems are needed to ensure that the right information is in the right hands in the appropriate format in time to act effectively.

The motivation for building effective intelligence systems must come from a shift in thinking about the global market and the profit opportunities of participating fully in that market. Implementing such a system and, more importantly, ensuring effective use of the results require fundamental changes in the total business system. It must include clear incentives to use the information and to devote the resources necessary to effect action. The exercise is not trivial, but it has become absolutely essential to long-term success in the international manufacturing environment.

4. Tap international capabilities by participating in global networks

Few manufacturing companies have the resources in-house to develop every promising technology, counter every competitive threat, and build a dominant presence in every important market. Success on a global scale requires that internal resources be used as effectively as possible and that they be leveraged with skills and resources available outside the firm. Building and participating in global networks can be an effective mechanism for gaining the most from internal capabilities while gaining access to external sources of competitiveness.

Three types of networks deserve attention. The first is simply networks between customers and suppliers. Collaborative relationships with competitors is the second type of network becoming more prominent in the international arena. Examples include joint research with foreign companies, contracting with other firms to do manufacturing, and joint product development and manufacturing with marketing remaining distinct. The third type, too often overlooked, is the network established among subsidiaries and divisions within the same firm. Companies' management of their participation in these types of networks has become a significant factor in their international competitive success.

Little is new, of course, in the concept of networks as a way to leverage resources, but the internationalization process is introducing new considerations that are not yet fully recognized by U.S. managers. For instance, close relations between suppliers and customers are an absolute necessity in the manufacture of many complex products. In some cases, foreign suppliers are the only competitive source of key components or equipment. With a few prominent exceptions, however, the concept of supplier relations as networks of technological capabilities and market information to be exploited in ways that go beyond simple contractual obligations has not been extensively embraced by U.S. manufacturers. Particularly with purchasing from foreign manufacturers, U.S. companies too often are motivated solely by the low cost of the product; the possibility of using the relationship to absorb key lessons about why the supplier's costs are lower, quality higher, and delivery times shorter than in-house production is simply ignored.

Collaborations with competitors create another type of network that, while essential in many businesses, can be mismanaged if the goals of the collaboration are not clearly delineated and the lessons learned not absorbed effectively into the organization. Managers must remember that the assets each partner contributes to such ventures depreciate over time; the market context in which the venture operates also varies, so the value of the venture to each partner will change over time. The dynamic nature of such relationships must be explicitly recognized. If collaborations with competitors are to be effective in building international networks and core competitive capabilities, managers must invest in the assets and activities that will sustain and maximize the value of the collaboration.[3]

Effective participation in international networks requires new managerial skills and explicit consideration of the full capabilities available in-house, including all foreign subsidiaries, and the trade-offs involved in sharing technology or market knowledge with competitors or potential competitors. In this context, using key competitive capabilities to obtain unique advantages from a shared source of information becomes especially important to competitive success. An organization's ability to learn through its international networks and to act on the resulting lessons is crucial in determining the nature and success of its global activities.

5. Maximize value

Achieving a combination of cost, quality, delivery, and performance that maximizes value to the customer has become the determinant of global manufacturing success. Maximizing value to the customer, for instance by mobilizing unique manufacturing capabilities to provide distinctive product features and to facilitate responsiveness, is the strategy most likely to build global market share for U.S. manufacturers. Moreover, a value-based strategy often provides a sustainable competitive advantage by building customer loyalty and capturing the benefits of well-developed competitive capabilities.

Improvements in production systems, fostered primarily by Japanese manufacturers, are defining the steps necessary to maximize value. Improving worker skills, redefining supplier-customer relations, integrating product and process design and engineering, using appropriate advanced

---

[3] Gary Hamel, Yves L. Doz, and C. K. Prahalad, "Collaborate with Your Competitors—and Win," *Harvard Business Review*, January-February, 1989, pp. 133-139.

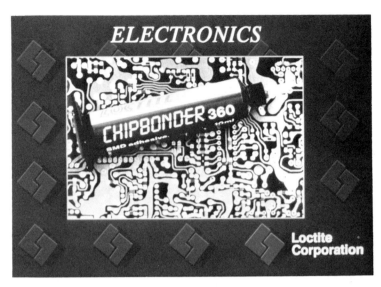

The quality aggression strategy of the Loctite Corporation has given the firm a major position in the growing world-wide electronics industry. The firm's Chipbonder TM 360, which permits surface mounted devices to be bonded to printed circuit boards during the manufacturing process, was developed to meet a specific need in the manufacture of electronic equipment.

manufacturing technologies, and creating information systems that convey the status of markets, competition, technology, and the total production system are keys to creating world-class manufacturing capabilities. For example, companies in a variety of markets are using their manufacturing systems to make the concept of quality aggression a central feature of their competitive strategies. These firms use technology and total quality control practices to command premium prices and strong customer loyalty as well as to reveal and minimize the costs of quality to strengthen profit margins, creating resources for innovation. Loctite Corporation, for one, has used a quality aggression strategy to become a leading producer of engineering adhesives and other specialty chemicals for world markets based on quality products and continuous advances in chemical technology. Similarly, Millipore has built a global business in fluid analysis and purification systems based on superior technology.[4]

An important point to bear in mind from a national perspective is that high-value products are not limited to relatively expensive, high-technology

---

[4] American Business Conference, *Winning in the World Market*.

products that succeed because of their sophistication. Although high-technology products will continue to be a source of competitive advantage for this country, standard technology products for commodity markets that compete on price and features also provide high value. U.S. manufacturers cannot afford to abandon these market segments. Continued participation depends on companies building effective manufacturing systems in their domestic plants and, at the same time, integrating the capabilities of foreign suppliers and partners in ways that strengthen core capabilities in manufacturing. Meeting these requirements represents the fundamental challenge of internationalization and the key to maximizing value in world markets.

6. Speed technology commercialization

A number of aspects of the internationalization process have combined to make the ability to commercialize new technologies quickly critical to success. The increasing number of foreign competitors with advanced technological capabilities has multiplied the potential sources of new products. Capital costs in some industries have made quick penetration of markets to gain production volume a necessity for profitability. Finally, short product life cycles in a growing number of industries are creating advantages for firms who are first to market.

In the current international manufacturing environment, managing the technology commercialization process requires an ability to be both a technology pioneer and a quick follower. U.S. companies have long viewed their ability to introduce new technologies into mass markets as a critical competitive advantage. In fact, U.S. corporations, universities, and government laboratories devote so many resources to the creation of new science and technology that pioneering new technologies must retain a central role in corporate strategy.

At the same time, the international mobility of technology and the technological sophistication of foreign competitors require an ability to be a rapid follower as well as a technology pioneer. U.S. managers too often overlook the importance of quickly matching the technological introductions of competitors, often preferring to take a leapfrog approach. Yet while the next breakthrough is being prepared, the U.S. firm misses the valuable learning experience of having a product competing in the marketplace and cannot take advantage of gradual technology iterations incorporated into the competitor's product. Leapfrogging, in effect, becomes more difficult when it involves a moving target.

Speeding technology commercialization is closely linked to all the other success factors identified. An essential aspect is the emphasis on global intelligence and the ability to feed the resulting information into an effective manufacturing organization. The ability to speed commercialization, even to make rapid product commercialization that maximizes customer value a core competitive capability, hinges on the intellectual capital available to U.S. manufacturers. Maximizing intellectual capital has been the implicit strategy of the United States in world markets, but there is disturbing evidence that this resource is slipping. The importance of developing and harnessing skills, expertise, and creativity embodied in intellectual capital is so great that it is the focus of the following suggestions for ensuring national success.

SOURCES OF NATIONAL SUCCESS

A central consequence of internationalization is that manufacturers have an unprecedented and growing degree of choice in their decisions on production and investment locations. As companies increase their ability to separate design, engineering, production, and marketing geographically while still maintaining a globally integrated manufacturing system, opportunities will grow to enjoy the benefits of participating in the U.S. market regardless of where production takes place. It is presumptuous to assume that market size alone will stimulate investments in high value-added manufacturing activities in this country. The attractiveness of a location for manufacturing investments is determined by many additional factors— low costs, a strong supplier base, well-trained workers, skilled engineers and managers, and access to advanced technology. The realities of global competition are such that the interests of manufacturing corporations— performing functions in their most effective location—may diverge from the national interest of attracting high-value manufacturing activities.

Because internationalization has created unprecedented competition among countries for manufacturing investments, a conscious effort is needed to build and leverage the national assets that will spur manufacturing activity in the United States. This country must have a comprehensive set of national policies that will ensure competitiveness with rapidly advancing foreign locations—policies that support a U.S. industrial base with the ability to attack global markets rather than policies that defend historical positions. This fundamental requirement suggests a number of national success factors. They focus on building national intellectual capital by spurring competition, investment, and knowledge creation.

1. Maintain open markets

It is not necessary to rehash the advantages of fair and open trade, but it is worth noting how internationalization is changing the dynamics of trade protection. In classic (and simple) terms, trade restraints represent a victory of producers beset by foreign competition over consumers of foreign goods. With the growth of international networks, however, the distinctions between domestic and foreign producers and between producers and consumers have blurred. In automobiles, for instance, the production networks that have emerged among U.S., Japanese, European, and Korean manufacturers in both parts and final products now make any type of trade restrictions difficult to implement effectively. The three U.S. producers have become customers, suppliers, and partners of foreign firms; they have built extensive manufacturing capacity abroad, especially in Mexico; and the major Japanese producers have established significant North American production capacity. Under these conditions the costs and benefits of protection would be difficult to assess. Similar developments in other industries, ranging from machine tools to electronics, have effectively precluded trade protection as a realistic option for helping U.S. producers.[5]

The argument for trade protection has taken a more aggressive tone in recent years. Many analysts claim that U.S. firms suffer from foreign competition in the domestic market without having equal access to competitors' markets. Though protecting the U.S. market *might* provide additional leverage in trade negotiations, there is strong evidence that such protection only temporarily affects foreign producers' U.S. market share as they increase U.S. production capacity. Current policy emphasis on opening foreign markets to U.S. exports and investment should be continued, but using domestic protection as a bargaining chip unjustifiably damages both consumers and producers.

Particularly given the futility of unambiguously capturing any benefits from trade protection, there is clearly no substitute for the level and strength of competition provided by open markets. International trade introduces new technologies, new management practices, and new strategic priorities that would otherwise penetrate U.S. industry slowly at best. Renewed attention to quality, efforts to improve worker participation in production management, changes in customer-supplier relationships, and other trends sweeping the manufacturing base have been forced by foreign competition. They have become the benchmarks of world-class manufacturing and are the clearest possible indication of the absolute necessity of open markets, both in the United States and abroad, to the long-term success of U.S. manufacturing.

2. Build intellectual assets

The competitive imperatives emerging in the international manufacturing environment are placing a premium on intellectual assets. In many

---

[5] The effects of international investment and alliances on trade policy, and the effects of trade policy on national competitiveness, are discussed in detail by Jagdish Bhagwati in *Protectionism*, Boston, Mass., MIT Press, 1988.

industries the emphasis has shifted from minimizing labor content and cost to attracting and using a skilled work force to gain competitive advantage from superior operation and maintenance of advanced manufacturing systems. If U.S. manufacturers are to use such advanced systems to provide superior value to global customers, and if foreign manufacturers are to perceive advantages in establishing high-value activities in this country, the United States must provide the intellectual capital to make such strategies profitable. That means continued investment in basic research and greater attention to education.

A great deal of effort has gone into improving primary and secondary education, and progress is being made on a local basis, but the overall quality of U.S. education remains too low. A U.S. aerospace firm, for example, discovered in a recent survey that its factory employees read on average at an eighth-grade level; math skills were only slightly better. The Office of Technology Assessment reports that U.S. twelfth graders ranked twelfth and fourteenth in geometry and algebra, respectively, among students in 15 developed and developing countries.[6] U.S. manufacturers cannot succeed when they must devote undue attention to product and process design to ensure quality production by a low-skilled work force. The situation can only get worse as international competition demands greater flexibility, more worker responsibility, and more use of sophisticated technology.

An often overlooked fact is that the people who will be working in U.S. manufacturing in the year 2000 are already in the work force. Improvements in national intellectual assets must include a focus on building the infrastructure to facilitate continuing education and retraining. Two- and 4-year technical colleges play a key role, but to be effective they must have a close relationship with industry. Because industry itself is often the source of new technologies and techniques, technical colleges can be important links in disseminating the best practices and upgrading the skills and knowledge of the nation's work force. Without dramatic improvements in U.S. education, admittedly a long-term process, the educated work forces in foreign countries will provide important competitive advantages and increase the attraction of those locations for new manufacturing investments by U.S. firms.

Although retraining and continuing education deserve emphasis, the need to strengthen intellectual assets also has ramifications for academe, especially U.S. engineering and business schools. Shortcomings in engineering education result in poor engineering practice in U.S. manufacturing corporations. Some analysts attribute cost advantages of foreign manufacturers

---

[6] Office of Technology Assessment, *Making Things Better: Competing in Manufacturing*, Washington, D.C., U.S. Government Printing Office, 1990, p. 13.

to their product designs rather than to any great differences in technology or input costs. For instance, an automobile door produced in the United States can cost twice as much as one produced in Japan, mainly because of design. Japanese engineers design for less scrap and fewer components and specify less expensive materials and more realistic performance margins.[7] Though some of these factors are a function of corporate culture and practice, much depends on how engineers are trained.

3. Reassess information requirements

The inability of national statistics to capture the full extent and meaning of U.S. integration with the international economy imposes increasing handicaps on the policymaking process. The rise of global competitors, the integration of production systems and corporate ownership worldwide, the inflow of direct investment into the United States, and the dispersion of technological excellence globally have been so extensive that U.S. policy and data gathering efforts have not kept pace. For example, in a recent study of U.S. merchandise trade statistics, the U.S. General Accounting Office found that monthly trade balances are highly volatile and do not necessarily indicate changes in trade performance, that the accuracy of export statistics is questionable, and that existing data do not reflect changes in global production wrought by internationalization.[8]

Though improvements in trade data are essential, attention should be paid to broader issues:

1. What information is needed to direct, participate in, and respond to developments in the international production system?
2. What mechanisms are available or need to be developed to acquire and disseminate such information in a timely manner?
3. What definitions and methodologies need to be established to provide an appropriate basis for analyzing the information to provide useful policy insight?

Lack of resolution of these questions will continue to create impediments to consistent policy development and will cause unnecessary political disputes with U.S. trading partners.

As a key example, consider the disputes that have already arisen over foreign investment in U.S. high-technology companies and the exclusion of foreign companies from government-sponsored research consortia both here and abroad. The lack of clear criteria for defining a manufacturer

---

[7] Leif G. Soderberg, "Facing Up to the Engineering Gap," *The McKinsey Quarterly*, Spring 1989, pp. 2-18.

[8] General Accounting Office, *Merchandise Trade Statistics: Some Observations*, Washington, D.C., U.S. Government Printing Office, April 1989.

as U.S. or foreign makes such discriminating policies appear capricious. Possible criteria for U.S. firms that could be consistently applied and internationally accepted include a specified level of activity, in all functions, to be performed within U.S. borders, and a majority equity interest to be held by U.S. citizens. Alternatively, the question could be made moot by negotiating to make foreign economic systems so similar to the U.S. system, in terms of openness, competition, procurement, and general government support, that questions of fairness are eliminated and nationality becomes a neutral issue. Current policy seems to be an amalgam of both of these approaches. A clearer definition of U.S. firms would provide a basis for including foreign interests in domestic policy debates and for reconciling national interests with global production regardless of nationality. It would also help to eliminate contradictory policies and provide a basis for establishing the benefits of foreign investment and foreign production in the United States.[9]

Defining corporate nationality, or determining that such definition is unnecessary, is one step in providing the policy basis for judging information requirements. The goal should be to gather sufficient appropriate data, from both domestic and foreign sources, to support policy needs and to provide the ability to anticipate and guide events in international manufacturing to a greater extent than is possible now.

4. Retain high value-added manufacturing as a key national competitive capability by providing a favorable manufacturing environment

The question of what manufacturing capabilities are essential to national well-being tends to arise in policy discussions only in the context of national defense. Those discussions ought to be broadened to encompass key issues of wealth creation, employment generation, and continued national ability to participate in important emerging market segments. Policymakers should recognize, however, that different industries and different manufacturing activities generate very different income and employment effects, which are impossible to predict a priori. It is, therefore, impossible to judge the full value of retaining or the cost of losing any particular manufacturing capability. Instead, the essential role of government should be to provide the incentive structure and resources necessary to encourage private manufacturers to perform high-value functions in the United States.

Assuming a national consensus that high-value manufacturing must remain a national competitive capability, what steps are necessary to create a favorable environment for such activities in the United States? Many have been suggested elsewhere. They include providing incentives for

---

[9] The issue of defining a U.S. company is discussed more fully by Robert Reich in "Who Is Us?," *Harvard Business Review*, January-February 1990, pp. 53-64.

long-term investment through favorable tax treatment of long-term capital gains, making the R&D tax credit permanent, revising antitrust restrictions to allow joint production agreements in addition to R&D consortia, providing additional incentives for U.S. students to pursue degrees in science and engineering, and strengthening global intellectual property rights. Although the costs and benefits of these and similar policy proposals are hotly debated, their assumed objectives of increasing investment, strengthening technological resources, and building intellectual capital are central elements in making the United States an attractive location for high-value manufacturing activities.

This economy-wide approach to ensuring the availability of human and capital resources needed for high-value manufacturing should be the focus of government policy. Inevitably, however, concern for the health of key industries will spark government intervention to ensure that certain capabilities, deemed necessary for national defense, for instance, remain in U.S. hands. Such debates are becoming more common now that the United States is no longer superior in a range of technologies from semiconductors to engineered materials. Government responses have included research support (Sematech, NCMS), trade protection (semiconductors, machine tools), and direct negotiation of technology transfer (FSX). These and other cases entail unique circumstances—for instance, different levels of foreign participation in the relevant U.S. industries and varied political importance. Nevertheless, when government intervention is deemed desirable, the decision should be based on thorough analysis of a consistent set of factors and the type of government help provided should be evaluated against all available options.[10]

The objective of retaining high-value manufacturing as a key national competitive capability implies a number of factors that should be considered in policy debates about assistance for industry. To reiterate the most important factors, the value of specific manufacturing capabilities should be defined not only in terms of criticality to defense systems but also in relation to technology and knowledge content, importance as a supplier to other industries, and importance to U.S. exports. Although job creation or protection is a major political motivator, potential employment effects need to be assessed in terms of the quality and knowledge content of the affected jobs, not just quantity.

When direct government intervention is deemed necessary to retain specific manufacturing capabilities, it should be done in a way that least distorts trade. Given the pervasiveness of the internationalization process,

---

[10]This discussion is not intended to imply that extensive analysis is not now performed before government intervention. See, for example, Congressional Budget Office. *The Benefits and Risks of Federal Funding for Sematech*, Washington, D.C., U.S. Congress, 1987.

strict trade protection sacrifices access to foreign technology, invites retaliation and thereby closes major foreign markets to U.S. exports, and diminishes the incentives for constant improvement in product and processes created by foreign competition. Other mechanisms, such as support for R&D, encouragement of consortia activities, direct subsidies for adoption of new technology, and public funding for development of technical standards, would be preferable to the indirect subsidies implied by trade protection.

In the context of high value-added manufacturing as a core competitive capability, it might be useful for policymakers to think of the national manufacturing base as if it were a single corporation. The United States must be able to interact with the international manufacturing community from a position of strength. With appropriate national policies to spur investment and research and to improve education, both foreign and domestic manufacturers will have incentives to perform high-value activities in the United States. Government policy should continue to emphasize both strengthening the national skill base and ensuring access to foreign research so that new technologies, regardless of origin, can be rapidly incorporated into U.S. products. Much as corporations will participate in international networks to help strengthen and exploit their competitive capabilities, the United States should take part in the interdependent global economy in the context of a strong domestic manufacturing and technology base with the capability to participate in high-value markets. With the internationalization process accelerating, a policy mix that builds core national competitive capabilities in high-value manufacturing, interacting with the global market to exploit and strengthen them, is the only sure way to achieve the ultimate national goal—to advance the national standard of living.

5
Conclusion

Internationalization is creating fundamental changes in the nature of the U.S. manufacturing base and the criteria for success in manufacturing businesses:

- The domestic market is shrinking relative to total world demand.
- Technology sources have multiplied in number and sophistication around the world.
- Cost priorities have shifted as labor content declines and manufacturing systems emphasizing total quality control become essential for world-class production.
- Political changes fostered by both governmental initiatives and increasing business integration are forcing critical adjustments in global manufacturing strategy.

In certain ways, U.S. companies have an advantage in this shifting landscape. They have a greater production presence in foreign markets than non-U.S. firms and have the largest domestic market on which to build the basis for a global assault. On the other hand, that domestic market has fostered a provincial attitude that has made U.S. manufacturers less interested in foreign markets and less in touch with technological developments globally than they might otherwise have been. Though U.S. multinationals have retained their global competitiveness very well, as measured by world market share, the bulk of domestic manufacturers face a range of new challenges as internationalization continues. Building the capabilities needed to prosper in the new manufacturing environment is a long-term process aimed at a moving target. What it takes to be competitive has changed and will change ever more rapidly, placing a premium on strengthening core capabilities and emphasizing learning and adaptability.

Pursuing such efforts at the national level can assure the success of the nation in the midst of the internationalization process. More than ever before, manufacturers have a choice of where to perform high-value functions because the requisite resources are available globally. Both implicitly and explicitly, countries compete for manufacturing investment. Creating an attractive, superior environment for conducting high-value manufacturing activities in the United States is the long-term key to national competitiveness. That requires a strong industrial infrastructure, including favorable conditions for long-term investment in R&D, technological innovation, and production; a strong supplier base; and well-educated workers, engineers, scientists, and managers. To maintain long-term advantages, policymakers must not only recognize the criticality of a strong infrastructure and the environment conducive to one but also must have appropriate information to anticipate, guide, and adapt to changes in the global market. With these prerequisites in place, retention and continued growth of high-value manufacturing as a national competitive capability should be assured.

Appendix
Indicators of Internationalization

A small sample of indicators yields substantial evidence of the extent of internationalization. A simple economic measure of the degree of internationalization of a nation's economy is the ratio of direct investment abroad to domestic wealth or assets or the ratio of assets or employment abroad to that at home. Using these ratios, Robert Lipsey found that internationalization has been growing over the past 20 years in six foreign countries, though all continue to lag the United States substantially.[1] Other economic measures show the same trend. For instance, between 1977 and 1988 the stock of foreign direct investment held by U.S. firms more than doubled, from $146 billion to $327 billion, of which more than 40 percent was in manufacturing. More importantly, the stock of foreign direct investment in the United States multiplied *15 times* between 1973 and 1988, reaching $328.8 billion; of that amount, $121.4 billion or about 37 percent was in manufacturing.[2]

These data deserve special emphasis. In the span of a decade a sea change occurred in global investment flows, with the United States shifting from the primary source country to the major home country. Beginning in 1978 the nation began to capture a growing proportion of foreign direct

---

[1] Robert E. Lipsey, "The Internationalization of Production," Working Paper No. 2923, National Bureau of Economic Research, Inc., Cambridge, Mass., 1989.

[2] U.S. Department of Commerce, Bureau of Economic Analysis. These numbers are based on historic costs rather than current value as represented by replacement costs.

investment inflows while supplying a decreasing share of outflows.[3] In 1985 the United States captured 40 percent of total foreign direct investment while supplying only 25 percent of outflows. By 1988 capital inflows into the United States reached $58.4 billion, of which $28.2 billion was in manufacturing.[4]

Another indicator of internationalization is the growing importance of trade in the U.S. economy. Both exports and imports virtually doubled as a percentage of GNP since 1970; exports are now about 13 percent of GNP and imports about 15 percent.[5] Manufactured exports as a percentage of shipments rose from 6.2 percent in 1978 to 8 percent in 1986; the fastest growth was in high-technology products, which rose from 25 percent of manufactured exports in 1975 to 37 percent in 1986.[6] As the importance of international trade has grown, so has the role of intrafirm trade. Trade carried out between affiliates of the same firm accounted for about 35 percent of U.S. exports and about 40 percent of imports in 1980. These percentages vary by industry and country, ranging as high as 80 percent of U.S. imports from Singapore.[7] For trade in manufactures alone, United Nations estimates for 1982 show intrafirm trade accounted for 39 percent of U.S. manufactured exports and 63 percent of manufactured imports.[8]

Another critical factor is the rise of nonequity forms of cooperation between international corporations, collectively called interfirm agreements, as major mechanisms for achieving internationalization. The use of licensing, cross-licensing, cooperative marketing and research agreements, organizational services, turnkey contracts, and other formal and informal arrangements between customers and suppliers, as well as between competing firms, has proliferated in the past decade. Though comprehensive data are difficult to obtain, various studies have documented the number and range of such alliances in a number of key industries.[9] For example, a recent study of international ventures in the semiconductor industry found 183 instances of international activity, including acquisitions and foreign direct investment, by firms based in the United States, Europe, Japan, and

---

[3] United Nations, *Transnational Corporations in World Development: An Overview*, 1988, p. 74.

[4] U.S. Department of Commerce, Bureau of Economic Analysis.

[5] Office of Technology Assessment, *Technology and the American Economic Transition*, Washington, D.C., U.S. Government Printing Office, 1988, p. 18.

[6] National Science Board, *Science and Engineering Indicators*—1989, pp. 376-377.

[7] Jane Sneddon Little, "Intra-Firm Trade and U.S. Protectionism: Thoughts Based on a Small Survey," *New England Economic Review*, January-February 1986, pp. 42-51.

[8] United Nations, *Transnational Corporations in World Development: An Overview*, p. 93.

[9] For example, see David C. Mowery, ed., *International Collaborative Ventures in U.S. Manufacturing*, Cambridge, Mass., Ballinger Publishing Co., 1988, and Vonortas, *The Changing Economic Context: Strategic Alliances Among Multinationals*.

Korea over the period 1978-1984; of these, 121 (66 percent) were interfirm agreements, including joint ventures, technology exchanges, licensing and cross-licensing, and second sourcing.[10] Studies of other industries, including machine tools, commercial aircraft, and automobiles, document major increases in the use of interfirm agreements and many cases in which such alliances have become the preferred route to internationalization.

Science and technology indicators also confirm the growing internationalization of the U.S. economy. Between 1970 and 1988 foreign applicants increased their share of U.S. patents awarded from about 27 percent to 48 percent. In particular, Japan's share quintupled, from 4 percent to 21 percent, over this period.[11] Another indicator is the number of foreign students in U.S. universities. In 1988 foreign students comprised approximately 46 percent of full-time enrollments in engineering graduate schools, received half of U.S. doctorates in engineering, and comprised 66 percent of engineering postdoctoral students.[12] This dominance of foreign students in the nation's universities is a major factor in the internationalization process. On one hand, they are an important resource for the United States since a significant number remain in this country. Those who go home, on the other hand, vastly upgrade the skills available in their countries, creating new opportunities for multinational manufacturers to perform R&D and engineering functions globally and to use sophisticated production technologies in plants worldwide.

These and other indicators clearly demonstrate both the growth of internationalization in the United States and the changing nature of the internationalization process. Though U.S. multinational corporations remain preeminent in global assets and have maintained their share of world manufacturing exports at about 18 percent,[13] foreign investment in the United States is increasing rapidly. Even more importantly, simple measures of assets no longer capture the pervasiveness of internationalization. The growing use of interfirm agreements in many forms is creating linkages among companies that are virtually impossible to quantify or even to document fully. These alliances are based on corporate assessments of strengths and weaknesses and strategic decisions on how best to leverage corporate resources to build the global market presence that is now crucial to competitive success.

---

[10] Vonortas, pp. 39-46.

[11] National Science Board, *Science and Engineering Indicators—1989*, p. 356.

[12] Ibid., pp. 218, 222, 225.

[13] Robert Lipsey and Irving Kravis, "The Competitiveness and Comparative Advantage of U.S. Multinationals, 1975-1983," Working Paper No. 2051, National Bureau of Economic Research, Inc., October 1986, Cambridge, Mass., p. 494.

Bibliography

Bale, Harvey E., Jr. 1985. The United States Policy Toward Inward Foreign Direct Investment. Vanderbilt Journal of Transnational Law 18(2):199-222.

Bartlett, Christopher A. 1989. Managing Across Borders: The Transnational Solution. Cambridge, Mass.: Harvard Business School Press.

Bhagwati, Jagdish. 1988. Protectionism. Boston, Mass., MIT Press.

Bergsten, C. Fred, Thomas Horst, and Theodore Moran. 1978. American Multinationals and American Interests. Washington, D.C.: The Brookings Institution.

Committee on Science, Engineering, and Public Policy, Panel on Technology and Employment. 1987. Technology and Employment: Innovation and Growth in the U.S. Economy. Washington, D.C.: National Academy Press.

Committee on Technology Issues that Impact International Competitiveness. 1988. The Technological Dimensions of International Competitiveness. Washington, D.C.: National Academy Press.

Cooney, Stephen. 1988. Manufacturing Creates America's Strength. Washington, D.C.: National Association of Manufacturers.

Creamer, D. 1976. Overseas Research and Development by United States Multinationals 1966-1975. New York: The Conference Board.

Dornbush, Rudiger, James Poterba, and Lawrence Summers. 1988. The Case for Manufacturing in America's Future. Rochester, N.Y.: Eastman Kodak Company.

Dunning, John H. 1982. International Production and the Multinational Enterprise. London: George Allen and Unwin.

Fosler, R. Scott. 1989. State Economic Policy: An Assessment. Business in the Contemporary World 1(4):86-97.

Fusfeld, Herbert I. 1989. Interaction Between National Policies and Technical Linkages of Multinational Enterprises. Troy, N.Y.: Center for Science and Technology Policy, Rensselaer Polytechnic Institute.

Graham, Edward M., and Paul R. Krugman. 1989. Foreign Direct Investment in the United States. Washington, D.C.: Institute for International Economics.

Grunwald, Joseph, and Kenneth Flamm. 1985. The Global Factory: Foreign Assembly in International Trade. Washington, D.C.: The Brookings Institution.

Guile, Bruce R., and Harvey Brooks. 1987. Technology and Global Industry—Companies and Nations in the World Economy. Washington, D.C.: National Academy Press.

Haklisch, Carmela S., and Nicholas Vonortas. 1987. Export Controls and the International Technical System: The U.S. Machine Tool Industry. Troy, N.Y.: Center for Science and Technology Policy, Rensselaer Polytechnic Institute.

Hamel, Gary, Yves L. Doz, and C. K. Prahalad. 1989. Collaborate with Your Competitors—and Win. Harvard Business Review (January-February):133-139.

Hayes, Robert H., Steven C. Wheelwright, and Kim B. Clark. 1988. Dynamic Manufacturing: Creating the Learning Organization. New York: The Free Press.

Helleiner, Gerald K. 1983. Intra-Firm Trade and the Developing Countries. New York: St. Martin's Press.

Howell, Thomas R., William Noellert, Janet MacLaughlin, and Alan Wm. Wolff. 1988. The Microelectronics Race: The Impact of Government Policy on International Competition. Boulder, Colo.: Westview Press.

Liner, Blaine, and Larry Ledebur. 1987. Foreign Direct Investment in the United States: A Governor's Guide. Washington, D.C.: National Governors' Association and the Urban Institute.

Lipsey, Robert E. 1988. Changing Patterns of International Investment in and by the United States. Cambridge, Mass.: National Bureau of Economic Research, Inc.

Lipsey, Robert E. 1989. The Internationalization of Production. Cambridge, Mass.: National Bureau of Economic Research, Inc.

Lipsey, Robert E., and Irving B. Kravis. 1987. The Competitiveness and Comparative Advantage of U.S. Multinationals 1957-1984. Banca Nazionale del Lavoro Quarterly Review 161 (June):147-165.

Lipsey, Robert E., and Irving B. Kravis. 1989. Technological Characteristics of Industries and the Competitiveness of the U.S. and Its Multinational Firms. Cambridge, Mass.: National Bureau of Economic Research, Inc.

Little, Jane S. 1986. Intrafirm Trade and U.S. Protectionism: Thoughts Based on a Small Survey. New England Economic Review (January/February):42-49.

Magaziner, Ira, and Mark Patinkin. 1989. The Silent War: Inside the Global Business Battles Shaping America's Future. New York: Random House.

Mansfield, Edwin. 1988. Industry Innovation in Japan and the United States. Science 241:1769-1775.

Mansfield, Edwin, and Anthony Romeo. 1980. Technology Transfer to Overseas Subsidiaries by U.S.-Based Firms. Quarterly Journal of Economics xcv(4):737-750.

McCauley, Robert, and Steven Zimmer. 1989. Explaining International Differences in the Cost of Capital. Federal Reserve Bank of New York Quarterly Review 14(2):7-28.

McKinnon, Ronald I. 1988. Monetary and Exchange Rate Policies for International Financial Stability: A Proposal. Journal of Economic Perspectives 2(1):83-103.

McKinsey & Company, Inc. 1987. Winning in the World Market. New York: American Business Conference, Inc.

Mowery, David C. 1987. Alliance Politics and Economics: Multinational Joint Ventures in Commercial Aircraft. Cambridge, Mass.: Ballinger Publishing Co.

Mowery, David C., ed. 1988. International Collaborative Ventures in U.S. Manufacturing. Cambridge, Mass.: Ballinger.

Mowery, David C., and Nathan Rosenberg. 1989. New Developments in U.S. Technology Policy: Implications for Competitiveness and International Trade Policy. California Management Review 32(1):107-124.

National Commission for Employment Policy. 1988. U.S. Employment in an International Economy. Washington, D.C.: National Commission for Employment Policy.

National Science Board. 1989. Science and Engineering Indicators—1989. Washington, D.C.: U.S. Government Printing Office.

Oppenheimer, Michael F., and Donna M. Tuths. 1987. Nontariff Barriers—The Effects on Corporate Strategy in High Technology Sectors. Boulder, Colo.: Westview Press.

Prahalad, C. K., and Gary Hamel. 1990. The Core Competence of the Corporation. Harvard Business Review (May/June)32(1):79-91.

Prahalad, C. K., and Yves L. Doz. 1987. The Multinational Mission: Balancing Local Demands and Global Vision. New York: The Free Press.

Prestowitz, Clyde V., Jr. 1988. Trading Places: How We Allowed Japan to Take the Lead. New York: Basic Books.
Reich, Robert B. 1987. The Rise of Techno-Nationalism. The Atlantic Monthly 259(May): 63-69.
Reich, Robert B. 1990. Who Is Us? Harvard Business Review (January-February):53-64.
Research and Policy Committee. 1986. Leadership for Dynamic State Economies. Washington, D.C.: Committee for Economic Development.
Schonberger, Richard J. 1986. World Class Manufacturing: The Lessons of Simplicity Applied. New York: The Free Press.
Shaiken, Harley, and Stephen Herzenberg. 1987. Automation and Global Production— Automobile Engine Production in Mexico, the United States, and Canada. La Jolla, Calif.: Center for U.S.-Mexican Studies, University of California, San Diego.
Skrzycki, Cindy. September 22, 1989. The Company as Educator: Firms Teach Workers to Read, Write. The Washington Post:G1.
Soderberg, Leif G. 1989. Facing Up to the Engineering Gap. The McKinsey Quarterly (Spring):2-18.
Spencer, Linda M. 1988. American Assets: An Examination of Foreign Investment in the United States. Arlington, Va.: Congressional Economic Leadership Institute.
Stokes, Bruce. June 20, 1987. Mexican Momentum. National Journal:1572-1578.
Stone, Charles F., and Isabel V. Sawhill. 1986. Labor Market Implications of the Growing Internationalization of the U.S. Economy. Washington, D.C.: National Commission for Employment Policy.
Tyson, Laura, William Dickens, and John Zysman, eds. 1988. The Dynamics of Trade and Employment. Cambridge, Mass.: Ballinger Publishing Co.
United Nations Centre on Transnational Corporations. 1988. Transnational Corporations in World Development. New York: United Nations.
U.S. Congress, Congressional Budget Office. 1987. The Benefits and Risks of Federal Funding for Sematech—A Special Study. Washington, D.C.: U.S. Government Printing Office.
U.S. Congress, Office of Technology Assessment. 1988. Technology and the American Economic Transition: Choices for the Future. Washington, D.C.: U.S. Government Printing Office.
U.S. Congress, Office of Technology Assessment. 1989a. Paying the Bill: Manufacturing and America's Trade Deficit. Washington, D.C.: U.S. Government Printing Office.
U.S. Congress, Office of Technology Assessment. 1989b. Statistical Needs for a Changing U.S. Economy. Washington, D.C.: U.S. Government Printing Office.
U.S. Congress, Office of Technology Assessment. 1990. Making Things Better: Competing in Manufacturing. Washington, D.C.: U.S. Government Printing Office.
U.S. Department of Commerce, Bureau of Economic Analysis. 1988. Foreign Direct Investment in the United States: Operations of U.S. Affiliates of Foreign Companies. Washington, D.C.: U.S. Department of Commerce.
U.S. General Accounting Office. 1989. Federal Statistics—Merchandise Trade Statistics: Some Observations. Washington, D.C.: U.S. Government Printing Office.
Vernon, Raymond. 1966. International Investment and International Trade in the Product Cycle. Quarterly Journal of Economics 80:190-207.
Vernon, Raymond. 1971. Sovereignty at Bay: The Multinational Spread of U.S. Enterprises. London: Longman.
Vernon, Raymond. 1977. Storm over the Multinationals: The Real Issues. New York: Macmillan.
Vonortas, Nicholas S. 1989. The Changing Economic Context: Strategic Alliances Among Multinationals. Troy, N.Y.: Center for Science and Technology Policy, Rensselaer Polytechnic Institute.
Wilkins, Mira. 1974. The Maturing of Multinational Enterprise: American Business Abroad from 1914 to 1970. Cambridge, Mass.: Harvard University Press.
The World Bank. 1989. World Development Report 1989. New York: Oxford University Press.

Theodore Lownik Library
Illinois Benedictine College
Lisle, Illinois